W0091493

BCL-2 Protein Family

ADVANCES IN EXPERIMENTAL MEDICINE AND BIOLOGY

Editorial Board:
NATHAN BACK, *State University of New York at Buffalo*
IRUN R. COHEN, *The Weizmann Institute of Science*
ABEL LAJTHA, *N.S. Kline Institute for Psychiatric Research*
JOHN D. LAMBRIS, *University of Pennsylvania*
RODOLFO PAOLETTI, *University of Milan*

A Continuation Order Plan is available for this series. A continuation order will bring delivery of each new volume immediately upon publication. Volumes are billed only upon actual shipment. For further information please contact the publisher.

BCL-2 Protein Family

Essential Regulators of Cell Death

Edited by

Claudio Hetz, PhD

Institute of Biomedical Sciences, FONDAP Center for Molecular Studies of the Cell (CEMC)
Millennium Nucleus for Neural Morphogenesis (NEMO), University of Chile
Santiago, Chile
and
Department of Immunology and Infectious Diseases, Harvard School of Public Health,
Boston, Massachusetts, USA

Springer Science+Business Media, LLC

Landes Bioscience

Springer Science+Business Media, LLC
Landes Bioscience

Copyright ©2010 Landes Bioscience and Springer Science+Business Media, LLC

All rights reserved.
No part of this book may be reproduced or transmitted in any form or by any means, electronic or mechanical, including photocopy, recording, or any information storage and retrieval system, without permission
in writing from the publisher, with the exception of any material supplied specifically for the purpose of
being entered and executed on a computer system; for exclusive use by the Purchaser of the work.

Printed in the USA.

Springer Science+Business Media, LLC, 233 Spring Street, New York, New York 10013, USA
http://www.springer.com

Please address all inquiries to the publishers:
Landes Bioscience, 1002 West Avenue, Austin, Texas 78701, USA
Phone: 512/ 637 6050; FAX: 512/ 637 6079
http://www.landesbioscience.com

The chapters in this book are available in the Madame Curie Bioscience Database.
http://www.landesbioscience.com/curie

BCL-2 Protein Family: Essential Regulators of Cell Death, edited by Claudio Hetz. Landes Bioscience
/ Springer Science+Business Media, LLC dual imprint / Springer series: Advances in Experimental
Medicine and Biology.

ISBN: 978-1-4419-6705-3

While the authors, editors and publisher believe that drug selection and dosage and the specifications and
usage of equipment and devices, as set forth in this book, are in accord with current recommendations
and practice at the time of publication, they make no warranty, expressed or implied, with respect to
material described in this book. In view of the ongoing research, equipment development, changes in
governmental regulations and the rapid accumulation of information relating to the biomedical sciences,
the reader is urged to carefully review and evaluate the information provided herein.

Library of Congress Cataloging-in-Publication Data

BCL-2 protein family : essential regulators of cell death / edited by Claudio Hetz.
     p. ; cm. -- (Advances in experimental medicine and biology ; v. 687)
  Includes bibliographical references and index.
  ISBN 978-1-4419-6705-3
  1. Apoptosis. 2. Tumor proteins. I. Hetz, Claudio. II. Series: Advances in experimental medicine and
biology, v. 687. 0065-2598 ;
  [DNLM: 1. Proto-Oncogene Proteins c-bcl-2--metabolism. 2. Proto-Oncogene Proteins c-bcl-2--
physiology. 3. Apoptosis--physiology. 4. Neoplasms--metabolism. 5. Neoplasms--pathology. W1
AD559 v.687 2010 / QU 55.2 B363 2010]
  QH671.B45 2010
  571.9'36--dc22
                            2010017334

FOREWORD

The BCL-2 family proteins, which have either pro- or anti-apoptotic activities, have been studied intensively for the past decade owing to their importance in the regulation of apoptosis, tumorigenesis and cellular responses to anti-cancer therapy. They control the point of no return for clonogenic cell survival and thereby affect tumorigenesis and host-pathogen interactions and regulate animal development. In this informative volume, Claudio Hetz has brought together experts in the apoptosis field to provide an overview of the emerging roles of BCL-2 family members in different physiological and pathological conditions. Several chapters focus on the role of the BCL-2 family in mitochondrial-dependent apoptosis as well as physiological processes in healthy cells such as metabolism, organelle morphogenesis and organelle stress; other chapters discuss biochemical and genetic mechanisms related to deregulation of apoptosis such as cancer. Authors include Anthony Letai and Owen O'Connor, expert in apoptosis and cancer; Luca Scorrano and Mariusz Karbowski, pioneers in identifying the crosstalk between the apoptosis machinery and mitochondrial dynamics; and Nika Danial, Hamsa Puthalakath, and Claudio Hetz, who have contributed to identifying novel functions of the BCL-2 family in different organelles ranging from metabolism to protein folding stress. An important focus of the book is considering the potential THERAPEUTIC benefits of targeting apoptosis pathways in the context of human disease. Readers interested in understanding how a cell handles stress and the consequences of dysregulation of this process for human disease will find this volume very valuable.

Richard J. Youle
Senior Investigator
National Institute of Neurological Disorders and Stroke
The National Institutes of Health
Bethesda, Maryland, USA

PREFACE

Apoptosis is a conserved cell death mechanism essential for normal development and tissue homeostasis. Although it presumably participates in the development of all cell lineages, aberrations in the expression of apoptosis-related proteins have been implicated in the initiation of a variety of human diseases including autoimmunity, immunodeficiency, cancer, neurodegenerative diseases and many others. Signaling pathways regulate the initiation of apoptosis through death receptors and the intrinsic mitochondrial pathway. The BCL-2 family of proteins represents a critical intracellular checkpoint in the apoptotic pathway where, notably, the founding member was first identified at the chromosomal breakpoint of the translocation (14;18) in human follicular B cell lymphoma. BCL-2 expression increases the resistance of cells to many different death stimuli, including cytokine withdrawal, DNA damage, endoplasmic reticulum stress and oxidative stress, among others.

The BCL-2 family of proteins is comprised of both pro- and anti-apoptotic members that are defined by the presence of up to four conserved α-helical BCL-2 homology domains. Each member of the BCL-2 family has distinct patterns of developmental expression, subcellular localization and differential responsiveness to specific death stimuli. The functional balance of pro- versus anti-apoptotic members at the mitochondrial membrane determines whether a cell will live or die. In addition, different BCL-2-related proteins are located in multiprotein complexes within other organelles such as the endoplasmic reticulum and nucleus, helping control diverse cellular processes beyond apoptosis. Recent evidence indicates that BCL-2 family proteins control glucose metabolism at the mitochondria in addition to mitochondrial dynamics and morphogenesis. These new findings highlight the physiologic relevance of the BCL-2 family in the integration of apoptosis with other homeostatic pathways.

In this book, scientists pioneering the field have compiled a series of focused chapters to highlight the relevance of the BCL-2 family of proteins in apoptosis, physiology and disease. Dr. Anthony Letai gives a comprehensive overview on the molecular regulation of apoptosis at the mitochondria and the hierarchical organization of the pathway. Together with Dr. Owen O'Connor, this basic knowledge is applied to discuss recent advances on targeting the core apoptosis pathway for the treatment of cancer. Drs. Lucca Scorrano and Mariusz Karbowski discuss molecular pathways emerging as regulators of mitochondrial dynamics and the contributions of such tightly controlled processes to disease conditions. To address alternative functions of BCL-2 family proteins, Dr. Nika Danial, Hamsa Puthalakath and myself describe novel functions beyond apoptosis at different organelles, including DNA damage, autophagy, glucose metabolism, protein folding and many others. Finally, Dr. Marie Hardwick summarizes alternative functions of the BCL-2 family in the brain, an emerging area of research with pathological relevance. In summary, this book constitutes an attempt to describe a fascinating area of research where physiology and biomedicine converge at different levels, revealing a trip from the molecular regulation of apoptosis to the impact of this process to the physiology of a whole organism.

Claudio Hetz, PhD
Institute of Biomedical Sciences,
FONDAP Center for Molecular Studies of the Cell (CEMC)
and
Millennium Nucleus for Neural Morphogenesis (NEMO)
University of Chile, Santiago, Chile
and
Department of Immunology and Infectious Diseases,
Harvard School of Public Health,
Boston, Massachusetts, USA

ABOUT THE EDITOR...

CLAUDIO HETZ received his BA in Biotechnology Engineering from the University of Chile in 2000. In his PhD work with Claudio Soto at Serono Pharmaceutical Research Institute, Geneva, he showed that Prion pathogenesis involves endoplasmic reticulum stress responses and apoptosis. In 2004 he joined as a postdoctoral fellow Stanley Korsmeyer's lab at Dana-Farber Cancer Institute, a pioneer in the apoptosis field. Together they discovered new functions of the BCL-2 family in organelle physiology. Dr. Hetz followed his projects in Laurie Glimcher's lab at Harvard. During this period he expanded his studies on neurodegeneration, addressing the connection between apoptosis and the unfolded protein response in vivo. Since 2007 he is an Assistant Professor at the University of Chile and adjunct professor at Harvard. His lab uses animal models to investigate the signaling responses involved in protein misfolding disorders and the role of the BCL-2 protein family in stress conditions. He was recently awarded with the TWAS-ROLAC Young Scientist Prize, also as finalist with the Eppendorf and Science Award in Neurobiology, and other important recognitions.

PARTICIPANTS

Benjamin Caballero
Institute of Biomedical Sciences
FONDAP Center for Molecular Studies
   of the Cell (CEMC)
and
Millennium Nucleus for Neural
   Morphogenesis (NEMO)
University of Chile
Santiago
Chile

Nika N. Danial
Department of Pathology
Harvard Medical School
and
Department of Cancer Biology
Dana-Farber Cancer Institute
Boston, Massachusetts
USA

Alfredo Gimenez-Cassina
Department of Pathology
Harvard Medical School
and
Department of Cancer Biology
Dana-Farber Cancer Institute
Boston, Massachusetts
USA

J. Marie Hardwick
Department of Molecular Microbiology
   and Immunology
Johns Hopkins University
Baltimore, Maryland
USA

Claudio Hetz
Institute of Biomedical Sciences
FONDAP Center for Molecular Studies
   of the Cell (CEMC)
and
Millennium Nucleus for Neural
   Morphogenesis (NEMO)
University of Chile
Santiago
Chile
and
Department of Immunology
   and Infectious Diseases
Harvard School of Public Health
Boston, Massachusetts
USA

Mariusz Karbowski
Biotechnology Institute
Medical Biotechnology Center
University of Maryland
Baltimore, Maryland
USA

Heather M. Lamb
Bloomberg School of Public Health
and
Bloomberg School of Public Health
Johns Hopkins University
Baltimore, Maryland
USA

Anthony Letai
Dana-Farber Cancer Institute
Harvard Medical School
Boston, Massachusetts
USA

Fernanda Lisbona
Institute of Biomedical Sciences
FONDAP Center for Molecular Studies
of the Cell (CEMC)
and
Millennium Nucleus for Neural
Morphogenesis (NEMO)
University of Chile
Santiago
Chile

Owen A. O'Connor
Herbert Irving Comprehensive Cancer
Center
College of Physicians and Surgeons
The New York Presbyterian Hospital
Columbia University
New York, New York
USA

Luca Paoluzzi
Herbert Irving Comprehensive Cancer
Center
College of Physicians and Surgeons
The New York Presbyterian Hospital
Columbia University
New York, New York
USA

Hamsa Puthalakath
Department of Biochemistry
School of Molecular Sciences
La Trobe University, Bundoora
Australia

Diego Rojas-Rivera
Institute of Biomedical Sciences
FONDAP Center for Molecular Studies
of the Cell (CEMC)
and
Millennium Nucleus for Neural
Morphogenesis (NEMO)
University of Chile
Santiago
Chile

Luca Scorrano
Dulbecco-Telethon Institute
Venetian Institute of Molecular Medicine
Padova
Italy
and
Department of Cell Physiology
and Metabolism
University of Geneva Medical School
Genève
Switzerland

Maria Eugenia Soriano
Dulbecco-Telethon Institute
Venetian Institute of Molecular Medicine
Padova
Italy

Daniel Tondera
Department of Pathology
Harvard Medical School
and
Department of Cancer Biology
Dana-Farber Cancer Institute
Boston, Massachusetts
USA

Thanh-Trang Vo
Dana-Farber Cancer Institute
Harvard Medical School
Boston, Massachusetts
USA

Ross T. Weston
Department of Biochemistry
School of Molecular Sciences
La Trobe University, Bundoora
Australia

Richard J. Youle
National Institute of Neurological
   Disorders and Stroke
The National Institutes of Health
Bethesda, Maryland
USA

Sebastian Zamorano
Institute of Biomedical Sciences
FONDAP Center for Molecular Studies
   of the Cell (CEMC)
and
Millennium Nucleus for Neural
   Morphogenesis (NEMO)
University of Chile
Santiago
Chile

CONTENTS

8. MITOCHONDRIA ON GUARD: ROLE OF MITOCHONDRIAL FUSION AND FISSION IN THE REGULATION OF APOPTOSIS... 131

Mariusz Karbowski

Chapter 1

Homeostatic Functions of BCL-2 Proteins beyond Apoptosis

Nika N. Danial,* Alfredo Gimenez-Cassina and Daniel Tondera

Abstract

Since its introduction in 1930 by physiologist Walter Bradford Cannon, the concept of homeostasis remains the cardinal tenet of biologic regulation. Cells have evolved a highly integrated network of control mechanisms, including positive and negative feedback loops, to safeguard homeostasis in face of a wide range of stimuli. Such control mechanisms ultimately orchestrate cell death, division and repair in a manner concordant with cellular energy and ionic balance to achieve proper biologic fitness. The interdependence of these homeostatic pathways is also evidenced by shared control points that decode intra- and extracellular cues into defined effector responses.

As critical control points of the intrinsic apoptotic pathway, the BCL-2 family of cell death regulators plays an important role in cellular homeostasis.[1-3] The different anti- and pro-apoptotic members of this family form a highly selective network of functional interactions that ultimately governs the permeabilization of the mitochondrial outer membrane and subsequent release of apoptogenic factors such as cytochrome c.[4] The advent of loss- and gain-of-function genetic models for the various BCL-2 family proteins has not only provided important insights into apoptosis mechanisms but also uncovered unanticipated roles for these proteins in other physiologic pathways beyond apoptosis (Fig. 1). Here, we turn our attention to these alternative cellular functions for BCL-2 proteins. We begin with a brief introduction of the cast of characters originally known for their capacity to regulate apoptosis and continue to highlight recent advances that have shaped and reshaped our views on their physiologic relevance in integration of apoptosis with other homeostatic pathways.

Introduction

The BCL-2 family is composed of anti- and pro-apoptotic proteins that share sequence homology within α helical segments known as BCL-2 homology (BH) domains. Pro-apoptotic BAX and BAK and all anti-apoptotic members of the family, such as BCL-2 and BCL-X_L, are "multi-domain" family members that share sequence homology within three to four BH domains (Fig. 2). The BH3-only pro-apoptotic proteins, such as BAD, BID, BIM, NOXA and PUMA, share homology within a single amphipathic BH segment, the BH3 domain, which is also known as the minimum death domain. BAX and BAK constitute a gateway to mitochondrial apoptosis in that their combined deletion affords resistance to all death stimuli that activate the intrinsic pathway of apoptosis.[5,6] BH3-only proteins are cell death initiators whose pro-apoptotic activity is latent unless activated. Post translational modifications such as phosphorylation, proteolytic processing and lipid modification of BH3-only proteins have emerged as possible

*Corresponding Author: Nika N. Danial—Department of Cancer Biology, Dana-Farber Cancer Institute, Boston, Massachusetts 02115, USA. Email: nika_danial@dfci.harvard.edu

BCL-2 Protein Family: Essential Regulators of Cell Death, edited by Claudio Hetz.
©2010 Landes Bioscience and Springer Science+Business Media.

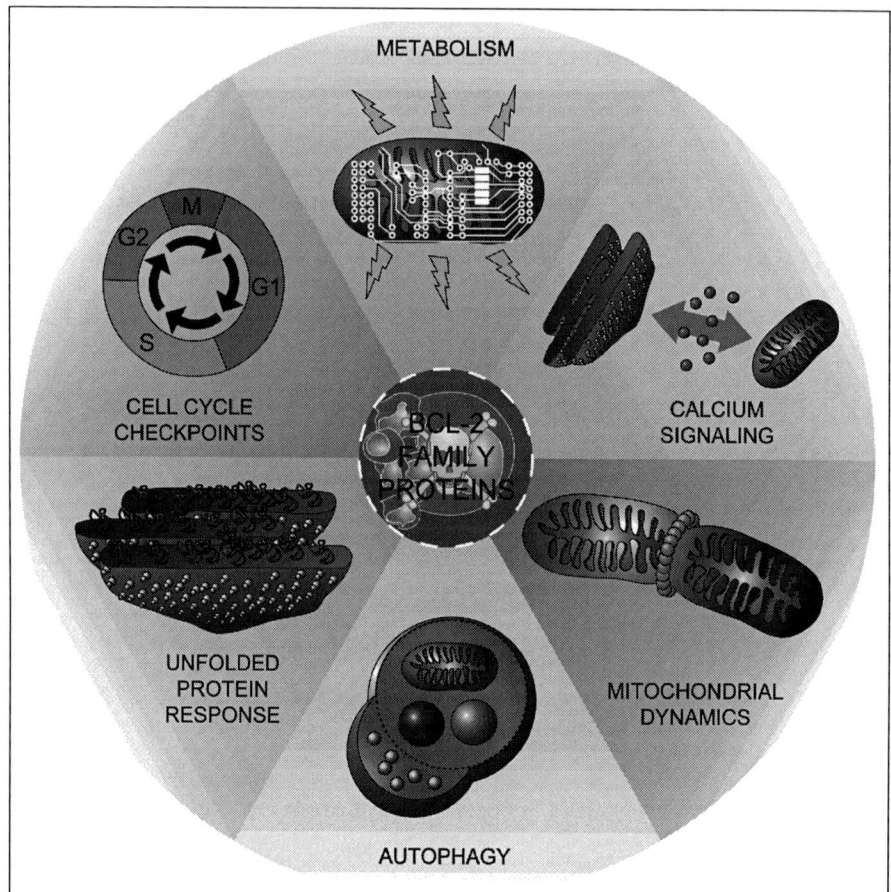

Figure 1. Physiologic pathways regulated by BCL-2 family proteins.

mechanisms that dynamically integrate extracellular survival and death signals with the core apoptotic machinery at the mitochondrion.[2,3] Pro-apoptotic activity of BH3-only proteins is further triggered by exposure of the hydrophobic face of their BH3 domain, facilitating interaction with a hydrophobic groove present in multi-domain partner proteins. This interaction can ultimately lead to BAX and BAK activation through both direct and indirect mechanisms. Structural studies and mutational analysis suggest that neutralization of anti-apoptotic molecules and direct activation of BAX and BAK are both necessary for the induction of mitochondrial outer membrane permeabiilizatin by apoptotic stimuli.[1,4]

While in-vitro studies demonstrate that over-expression of BH3-only molecules in a variety of cell lines leads to apoptosis, genetic loss-of-function models indicate that individual BH3-only proteins serve as cell death initiators responding to selected signals in restricted cell types. The lack of a universal defect in apoptosis upon deletion of a single BH3-only molecule may be explained in part by the functional redundancies among these pro-apoptotic proteins, the degree of which may vary depending on the specific cell type or death signal examined. The multiplicity of BH3-only proteins and restricted apoptosis defects associated with their ablation could also denote specialization, enabling the core apoptotic machinery to receive extensive input under different stress conditions, such as DNA damage, metabolic decline, growth fac-

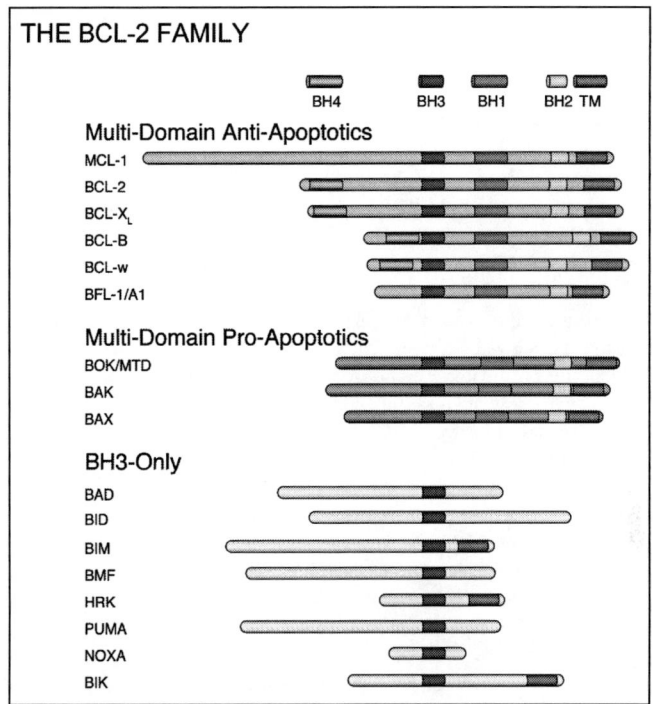

Figure 2. BCL-2 family proteins. BH: BCL-2 homology domain; TM: transmembrane domain.

tor withdrawal, hypoxic insult and unfolded proteins. Furthermore, such specialization may stem from active participation of these proteins in select physiologic pathways in the absence of death stimuli. Indeed, emerging biochemical and genetic evidence also support the notion that BCL-2 family members are integral component of distinct homeostatic pathways and carry out "day jobs" that can be distinguished from their capacity to regulate apoptosis.[2,7] By being nested in these physiologic pathways, they serve as critical checkpoints for death when cellular homeostasis is violated.

In the following sections, we explore biologic functions of BCL-2 proteins in mitochondrial energy metabolism, including metabolite transfer across the mitochondrial membrane, ER-mitochondrial trafficking of calcium, mitochondrial dynamics and nutrient metabolism. We further explore additional roles for these proteins in cell cycle checkpoints and cellular quality control mechanisms such as autophagy and unfolded protein response (Fig. 1). While some of these functions, such as mitochondrial dynamics and inter-organellar calcium trafficking, were uncovered during investigation of apoptotic mechanisms, their significance in healthy cells is being increasingly appreciated.

Mitochondrial Energy Metabolism

The striking discovery that many BCL-2 family proteins reside at mitochondria has long propelled investigation into their involvement in normal mitochondrial physiology. Beyond release of apoptogenic factors, mitochondria participate in calcium and metabolite trafficking, heme and steroid biosynthesis and serve as major source of cellular ATP. Below we highlight recent advances in elucidating the role of BCL-2 family proteins in calcium trafficking and metabolite transport across the mitochondria, which directly impact mitochondrial energy metabolism.

These processes are further influenced by mitochondrial ultrastructure and morphology that may additionally receive input from BCL-2 proteins.

Functional Interaction of BCL-X$_L$ and VDAC in Regulation of Mitochondrial Outer Membrane Permeability to Metabolites

The mitochondrial voltage-gated anion channel (VDAC) is the primary means of anion and metabolite transfer across the outer mitochondrial membrane. In its monomeric form, VDAC displays functionally defined ion selectivity, voltage gating and conductance.[8] Early reports on the association of select BCL-2 proteins with VDAC gave rise to several, often conflicting models, as to its involvement in apoptosis. While this has been greatly debated,[9] recent genetic studies ruled out any direct involvement of VDAC in mitochondrial apoptosis.[10] However, an accessory role for this protein in modulating the apoptotic response under select settings, including apoptosis induced by Ca^{2+}-mobilizing agents, is a viable possibility. Importantly, how the partnership between BCL-2 proteins and VDAC may influence metabolite flux across the outer mitochondrial membrane and mitochondrial energy metabolism is under active investigation. Below we highlight recent findings on the functional interaction between BCL-X$_L$ and VDAC and its potential physiologic role in shaping synaptic transmission.

Initial evidence in support of direct association of BCL-X$_L$ and VDAC emerged from independent biochemical studies using co-immunoprecipitation assays, gel filtration and chemical cross linking.[11,12] Recombinant BCL-X$_L$ was shown to increase the probability of VDAC's open configuration and its conductance in planar phospholipid membranes.[13] Importantly, this biophysical property of BCL-X$_L$ does not depend on its capacity to independently form ion channels in synthetic lipids. While addition of recombinant BCL-X$_L$ alone does not alter the open configuration of VDAC, it counters the effect of NADH, which normally promotes VDAC closure.[13,14] Recent structural characterization of BCL-X$_L$ in complex with VDAC has provided important insights into these biochemical interactions.[11,15] VDAC-1 adopts a β barrel structure composed of 19 strands in detergent micelles thought to closely mimic its three dimensional configuration in the lipid bilayer. The predominant interaction interface between VDAC-1 and BCL-X$_L$ is formed by strands 17 and 18 on VDAC and residues 94-156 in human BCL-X$_L$ mapping to helices 5 and 6 required for membrane insertion. Remarkably, the interaction interface of NADH and BCL-X$_L$ with VDAC share certain common residues, suggesting competition for binding. This may explain the dueling effect of NADH and BCL-X$_L$ in determining VDAC conductance.[11,13,15] The functional implications of these observations are especially intriguing in light of the proposed role for glycolytically-derived NADH in promoting VDAC closure and mediating the inhibitory effect of glycolysis on mitochondrial permeability to metabolites and respiration (also known as the Crabtree effect).[14] Whether by antagonizing NADH, BCL-X$_L$ overrides the inhibitory effect of glycolysis on mitochondrial respiration would be an interesting thesis for future experimentation.

It is likely that BCL-X$_L$ modulation of VDAC in-vivo is under complex regulatory mechanisms. This is especially relevant as beyond BCL-X$_L$, multiple protein and lipid interactions can influence VDAC, including hexokinase, the dynein light chain Tctex-1, the mitochondrial heat-shock protein 70 (mtHSP70), actin, tubulin and mitochondrial lipids such as cardiolipin and phosphatidylethanolamine.[16-21] Moreover, phosphorylation of VDAC by multiple kinases such as GSK3β, PKCε, C-RAF and PKA may influence VDAC's gating properties and its protein interaction network.[22-25] Phospho-regulatory mechanisms also influence the availability of BCL-X$_L$ for VDAC binding. For example, in its dephosphorylated state, the BH3-only protein BAD engages BCL-X$_L$ and precludes its association with VDAC.[26] Characterization of phospho-regulatory mechanisms impinging on VDAC biophysical properties will likely provide important insight into how transfer of metabolites across the mitochondrial outer membrane and mitochondrial respiration are coordinated with other cellular signaling pathways.

The capacity of BCL-X$_L$ to modulate the permeability of the mitochondrial outer membrane may have implications in ion homeostasis, especially in excitable cells such as neurons. The synaptic terminal in neurons is highly enriched in mitochondria, which actively participate in synaptic transmission by providing ATP and shaping the calcium transients during synaptic activity.[27] A specialized patch clamp technique applied to the giant squid presynaptic terminal enabled recordings of the mitochondrial outer membrane conductance for the first time in intact cells.[28] Inclusion of recombinant human BCL-X_L in the giant squid presynaptic terminal did not change the kinetics or amplitude of the presynaptic action potential but enhanced the rate of rise in postsynaptic response.[29] The molecular mechanism underlying the effect of BCL-X_L on synaptic transmission is not fully clear. One model proposes that increased synaptic activity in the presence of BCL-X_L may stem from its ability to facilitate ATP release from mitochondria, which is consistent with its binding to VDAC.[29] Recent evidence also suggests that the hydrophobic groove of BCL-X_L known to accommodate the BH3 domain of pro-apoptotic partners or BH3-mimetic compounds may modulate the effect of BCL-X_L on mitochondrial conductance.[30] Whether the biochemical properties that regulate BCL-X_L modulation of synaptic transmission and those in charge of its association with VDAC share common elements is not fully understood.

Deciphering Calcium Signals: BCL-2 Family and the ER-Mitochondria Cross Talk

Mitochondrial energy metabolism is tightly coordinated with intracellular Ca^{2+} signaling and mitochondrial Ca^{2+} content.[31] Mitochondrial Ca^{2+} uptake is facilitated by VDAC in the outer mitochondrial membrane and primarily mediated by a calcium uniporter in the inner mitochondrial membrane (Fig. 3A).[32] The efficiency of Ca^{2+} trafficking to mitochondrial matrix is ensured by an intricate inter-organellar architecture that physically links the ER and mitochondria over ~5% of their surface, providing high Ca^{2+} microdomains for efficient and rapid mitochondrial uptake of Ca^{2+} released by the ER.[33-35] The protein constituents and physical parameters of the molecular bridge between these two organelles are just being unraveled.[36-38] The ionic cross talk between the ER and mitochondria fine tunes mitochondrial energy metabolism with cellular Ca^{2+} signals under resting conditions as well as physiologic and pathophysiologic responses to Ca^{2+}-mobilizing agents.

The ER Ca^{2+} content is the net outcome of calcium uptake, release and leak through highly regulatable processes. Calcium is taken up by sarco/endoplasmic reticulum Ca-ATPase (SERCA) and released by inositol 1,4,5-trisphosphate (IP3) receptor or ryanodine receptor Ca^{2+} channels (Fig. 3A).[39] ER Ca^{2+} content is a chief determinant of the amount of Ca^{2+} that can be released in the cytosol and thus constitutes an important regulator of Ca^{2+} signals known to control a myriad of cellular functions, including survival and death.[39,40] The pattern of Ca^{2+} release from the ER, pulsatile (oscillatory) or sustained, may further influence these cellular responses to calcium.[41] Several member of the BCL-2 family localize to the ER and regulate ER Ca^{2+} content.[42-46] Genetic and pharmacologic approaches demonstrated that reduced ER Ca^{2+} content and inhibition of IP3R-mediated Ca^{2+} release is an important component of the pro-survival effect of anti-apoptotic BCL-2 proteins, which is reversed by the pro-apoptotic members of the family (Fig. 3b).[45,47-54] While these studies have helped unravel ER Ca^{2+} release as another important control point in apoptosis, several observations have implicated select anti-apoptotic BCL-2 proteins in active regulation of both resting and agonist-induced Ca^{2+} mobilization in healthy cells. In the following paragraphs, we limit our discussion to the proposed molecular mechanisms underlying regulation of ER Ca^{2+} content by BCL-2 family proteins and their potential implications in cellular physiology and energy metabolism beyond survival/death decisions. For a comprehensive treatment of pro-apoptotic Ca^{2+} signaling at the ER and its modulation by BCL-2 family proteins, the reader is referred to excellent recent reviews.[40,41,55,56]

Biochemical studies coupled with pharmacologic approaches have provided evidence that the effect of multi-domain anti- and pro-apoptotic BCL-2 proteins on ER Ca^{2+} is likely independent of their pore forming properties observed in synthetic membranes.[57] Rather, modulation of ER Ca^{2+} content by BCL-2 proteins is mediated, at least in part, by their direct or indirect modulation of IP3R[54,58,59] and SERCA.[60-62] The functional consequences of these interactions in regulation of ER Ca^{2+} handling and the underlying molecular mechanisms are not fully understood. Several scenarios have been proposed that are not always in agreement.[56,63] For example, BCL-2 and BCL-X_L associate

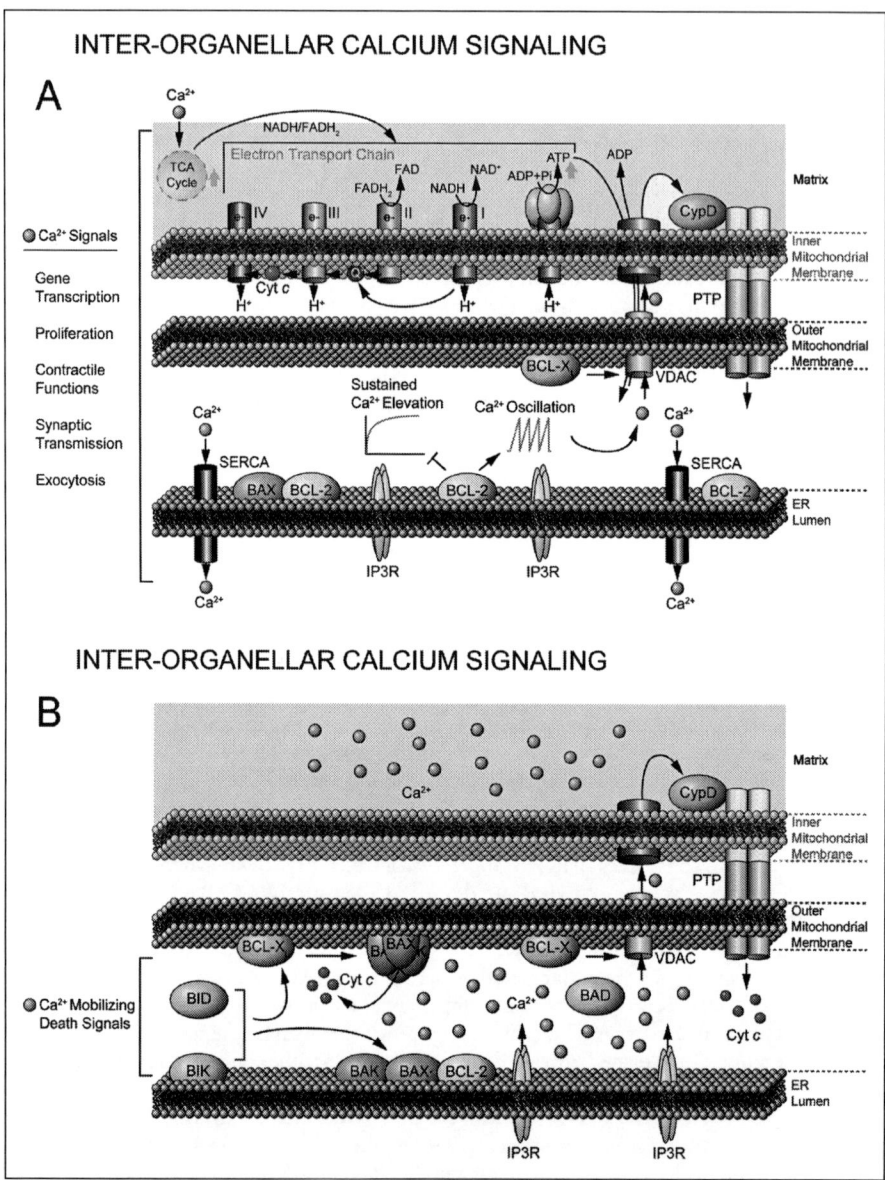

Figure 3. Calcium trafficking between ER and mitochondria in energy metabolism (A) and apoptosis (B). A) Calcium release from the ER stimulates energy metabolism in mitochondria. Through their effect on IP3R and VDAC, select members of the BCL-2 family modulate ER calcium release and mitochondrial calcium uptake, respectively. ER-mitochondrial calcium trafficking influences cytosolic calcium signals, which in turn impacts a myriad of calcium-dependent cellular functions, including transcription, proliferation, contractile functions, synaptic transmission and exocytosis. B) Exaggerated calcium released from the ER or taken up by the mitochondria switches the outcome of calcium signaling from energy metabolism to cell death through BAX/BAK dependent permeabilization of the mitochondrial outer membrane or opening of the mitochondrial permeability transition pore (PTP).

with IP3R[54,58,59] and may prevent the opening of the IP3R channel in synthetic lipid membranes according to some studies.[58] According to other reports, however, this association may enhance IP3R channel activity by sensitizing the channel to low concentration of IP3[54] or may cause Ca^{2+} leak from the ER.[59,64,65] The biochemical interaction of BCL-X$_L$ with the IP3R is opposed by pro-apoptotic proteins such as BID and BAX/BAK (Fig. 3B).[54,59] The potential regulatory effect of BCL-2 on IP3R channel activity may further involve phospho-regulatory mechanisms likely based on its ability to bind and recruit calcineurin to IP3R, favoring the dephosphorylated/inactive state of the channel.[59,66] BCL-2 itself is also subject to phosphorylation.[67] Indeed phosphorylated BCL-2 shows a predominant ER localization, potently induces Ca^{2+} release from IP3R channels and renders cells more susceptible to apoptosis.[68] Conversely, a nonphosphorylatable mutant of BCL-2 reduces ER Ca^{2+} content. BCL-2 was also shown to bind and inhibit SERCA1[60] and SERCA2b[62] likely by inducing their partial unfolding.[60] In addition, biochemical studies revealed that addition of recombinant BCL-2 to purified sarcoplasmic reticulum from rat muscle altered SERCA localization to specialized lipid microdomains.[61]

In light of specialized functions of the ER in sphingolipid and cholesterol metabolism, the effect of BCL-2 on IP3R and SERCA activity may well have implications beyond regulation of cellular survival/death. Furthermore, the capacity of ER to sort and regulate sub-cellular localization of newly synthesized proteins is tightly coupled with the luminal Ca^{2+} concentrations[69] and may be further attuned to BCL-2 family modulation of ER Ca^{2+} handling. Additional studies are required to fully explore possible roles for BCL-2 family proteins in modifying these specialized ER functions. Moreover, alterations in ER Ca^{2+} homeostasis can directly impact mitochondria. For example, BCL-2 and BCL-X$_L$ promote Ca^{2+} oscillations through the IP3R that are higher in frequency and longer in duration.[54,64,70] The oscillatory Ca^{2+} signals from the ER are sensed and decoded by mitochondria to enhance energy metabolism (Fig. 3A).[71] In addition to its effect on IP3R-mediated Ca^{2+} release and oscillations,[54] BCL-X$_L$ may influence ER-mitochondria Ca^{2+} trafficking at the level of mitochondrial Ca^{2+} uptake through its functional interaction with VDAC.[13,15] These observations warrant further investigation of the spatiotemporal mechanisms mediating BCL-X$_L$ effect on the ionic communication between ER and mitochondrial. As such, the possibility that BCL-X$_L$ may serve as active/functional component of the high Ca^{2+} microdomains linking mitochondria and ER is an intriguing thesis.

Several mitochondrial targets of Ca^{2+} have been identified and include metabolite carriers such as the aspirate/glutamate carrier[72] TCA cycle enzymes such as NAD^+-isocitrate, 2-oxoglutarate and pyruvate dehydrogenases[73,74] and the F0/F1-ATPase.[75] The integrated outcome of mitochondrial Ca^{2+} signaling manifests in enhanced TCA cycle activity, mitochondrial NADH, oxidative metabolism and ATP production (Fig. 3A).[71,74,76] However, supra-physiological levels of Ca^{2+} taken up by mitochondria can induce the opening of the permeability transition pore (PTP) and apoptosis.[77] Importantly, functional interaction of BCL-X$_L$ with VDAC can normally counter sensitization of PTP to Ca^{2+}.[78] Remarkably, dephosphorylation of the BH3-only protein BAD proved a physiologic mediator of ceramide-induced displacement of BCL-X$_L$ from VDAC and death associated with permeability transition.[26] While enhancement of mitochondrial bioenergetics in cells over-expressing BCL-2/ BCL-X$_L$ and inhibition of PTP sensitization to Ca^{2+} can protect cells from apoptosis, the broader implications of these findings may be homeostatic roles for BCL-2 family proteins in modulation of Ca^{2+} signals in healthy cells. For example, by controlling inter-organellar communication between the ER and mitochondria, BCL-2 family protein may help fine tune Ca^{2+} release and integrate cellular responses to Ca^{2+} signals such as gene transcription, proliferation, mitochondrial bioenergetics, redox capacity, contractile functions, synaptic transmission and exocytosis.[79]

Remodeling the Mitochondrial Membrane: BCL-2 Family Members and Mitochondrial Dynamics

The mitochondrial ultrastructure is highly complex and consists of two membranes; the inner and outer mitochondrial membranes. The inner mitochondrial membrane is invaginated into the matrix space to form the cristae; the sites of oxidative phosphorylation. Mitochondrial cristae are separated from the rest of the intermembrane space by narrow tubule-like structures known as cristae junctions ranging from 20-50 nm.[80] This exquisite structural design has provided major insights into the dynamic capacity of mitochondrial inner membrane to change the number and size of cristae junctions ultimately regulating ion and substrate diffusion for electron transfer reactions and oxidative phosphorylation, which take place at the mitochondrial inner membrane.[81] For example, heightened demand for oxidative phosphorylation prompts transition of mitochondrial ultrastructure from the so-called "orthodox" (expanded matrix) to the "condensed" state marked by compacted matrix and enlarged cristae. On the other hand, depletion of ADP during transition from respiratory state 3 to 4 is associated with remodeling of cristae from the condensed to the orthodox morphology.[82] In addition to the complex mitochondrial ultrastructure, global mitochondrial morphology effectively integrates mitochondrial shape/structure with function. Mitochondrial morphology is the byproduct of an intricate balance of mitochondrial fission and fusion events leading to punctate mitochondria or filamentous network, respectively.[83,84] Importantly, mitochondrial dynamics marks a quality control mechanism whereby damaged mitochondria selectively fuse with healthy mitochondria, exchange contents and undergo fission. Dysfunctional mitochondria that are not adequately repaired after this fusion and fission cycle are subsequently eliminated by autophagy.[85]

The protein components of the mitochondrial fission and fusion machinery are highly conserved between yeast and mammals.[83,84] Mitochondrial fission is governed by the large dynamin-like GTPase such as dynamin-related protein 1 (DRP1, Dnm1p in yeast), which shuttles between the cytosol and the mitochondrial outer membrane and the integral outer mitochondrial membrane protein FIS1 (Fig. 4).[86-92] According to the current model, FIS1 serves as an anchor for DRP1 binding to the outer mitochondrial membrane. In order to divide mitochondrial tubules, DRP1 then forms ring-like structures around the mitochondrion and facilitates its division (Fig. 4).[86-92] Mitochondrial fusion is executed by another set of large GTPases, Mitofusin (MFN) 1 and 2 (yeast Fzo1p) in the outer mitochondrial membrane and the optic atrophy 1 (OPA1) protein (yeast Mgm 1) in the inner mitochondrial membrane (Fig. 4). OPA1 regulates cristae fusion and remodeling, while mitofusins mediate tethering and subsequent fusion of adjacent mitochondria.[93-98]

The observation that apoptosis is accompanied by mitochondrial fragmentation provided the first indication of a cross talk between BCL-2 family proteins and mitochondrial dynamics. Below we briefly summarize the findings that initially integrated mitochondrial morphology with the cell death/survival function of BCL-2 proteins and subsequently highlight studies that uncovered functional interactions of these proteins and the mitochondrial dynamics machinery separate from their cell death regulatory function.

The mitochondrial network undergoes dramatic changes during apoptosis marked by fragmentation of individual units that cluster around the nucleus.[99] DRP1 was identified as a key factor in mitochondrial fragmentation during apoptosis.[100,101] Loss of DRP1's function delays or inhibits BAX/BAK dependent mitochondrial outer membrane permeabilization (MOMP).[100,102-105] BAX and BAK colocalize with DRP1 and MFN2 at mitochondrial fission sites and DRP1/BAX-containing foci inhibit MFN2-mediated fusion activity, suggesting a role for these pro-apoptotic molecules in mitochondrial fragmentation during apoptosis (Fig. 4).[101] Conversely, a constitutively active mutant of MFN2 has been shown to directly interfere with BAX activation induced by staurosporine and free radicals.[106] Recent biochemical studies provided evidence for BAX/BAK-dependent selective sumoylation of DRP1, which locks the cellular DRP1 pool to mitochondrial foci as opposed to its rapid cycling between the cytosol and mitochondria in steady state.[107] This posttranslational modification is concurrent with mitochondrial recruitment of BAX, but prior to cytochrome *c* release. The molecular underpinnings of this BAX/BAK-regulated sumoylation have not been fully explored. Importantly, whether and how these molecular transactions between BAX/BAK and DRP1 are influenced by BH3-only proteins is an intriguing thesis to be tested in future studies.

The collective findings from loss- and gain-of-function studies have suggested pro- and anti-apoptotic roles for mitochondrial fission and fusion proteins, respectively. Mitochondrial fragmentation

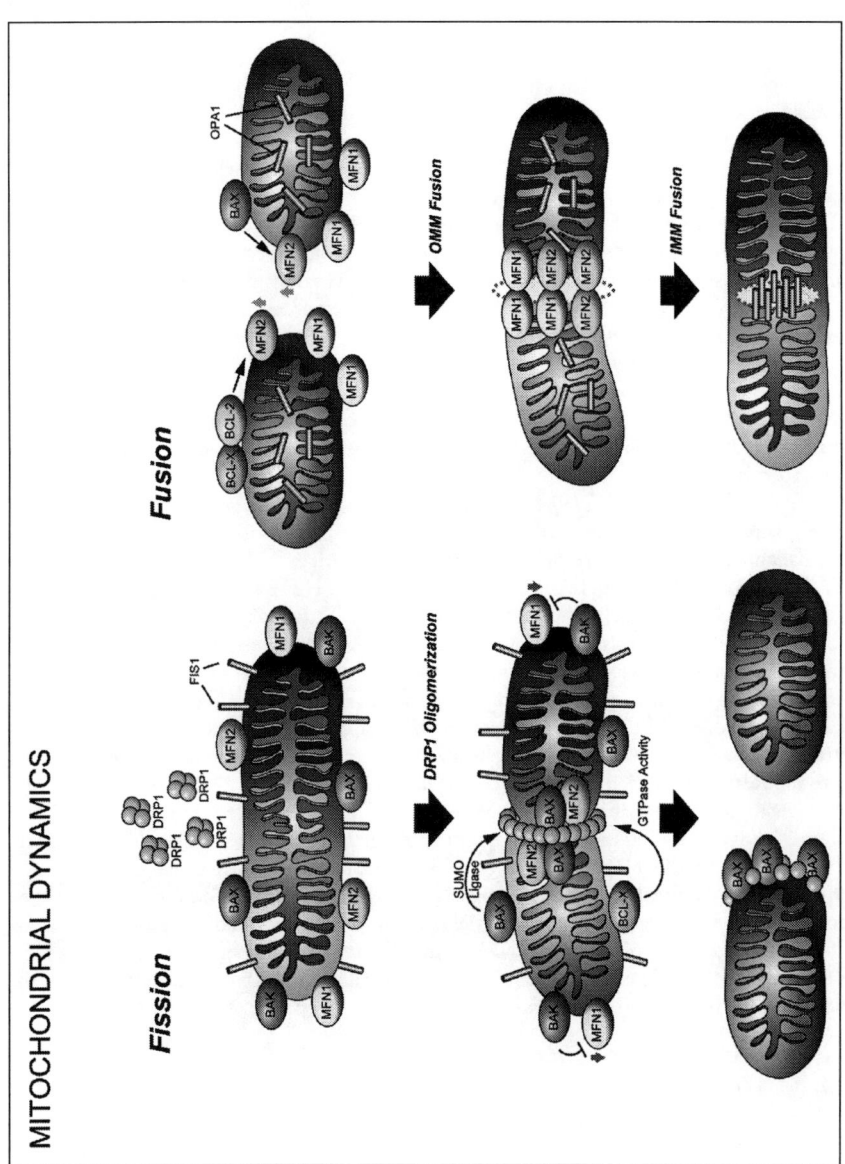

Figure 4. Modulation of mitochondrial fusion and fission by BCL-2 proteins.

is contemporaneous with BAK/BAK-mediated MOMP and contributes to release of cytochrome *c* from mitochondria. However, the mechanisms and consequences of mitochondrial dynamics in the regulation of cell death and survival remain to be fully elucidated. Importantly, mitochondrial fragmentation as such does not cause apoptosis.[108-111] For example, mitochondria fragment during mitosis so that they can be efficiently segregated to daughter cells.[112] Furthermore, reversible mitochondrial fission may regulate intracellular distribution/movement of this organelle, its calcium buffering capacity and intracellular calcium signals.[113]

BAX and BAK positively regulate mitochondrial fusion in healthy cells contrary to their effect in promoting mitochondrial fission during apoptosis.[114,115] In the absence of any death signal, cells doubly deficient for BAX and BAK have aberrant mitochondrial morphology marked by a high degree of fragmentation that can be genetically corrected by reintroduction of either BAX or BAK, MFN2, but not by a DRP1 loss-of-function mutant.[115] Interestingly, ablation of BAX and BAK alters the distribution pattern of MFN2 at the mitochondria and the size of higher order mitochondria-resident protein complexes containing MFN2, which could be also corrected by reintroduction of BAX in double knockout cells. It has also been suggested that BAX or BAK may regulate the GTPase activity of MFN2 known to be required for its mitochondrial fusion activity (Fig. 4).[115] Thus, BAX and BAK appear to have opposite effects on mitochondrial morphology in healthy cells versus those committed to undergo apoptosis. The molecular mechanisms underlying this functional dichotomy and transition from pro-fusion to pro-fission effect by the same proteins are not fully known. However, both programs require the BH3 domain.[115] In addition, the functional consequences of the fusion-promoting role of BAX/BAK in healthy cells remains to be fully explored, including their precise effect on mitochondrial bioenergetics separate from their pro-apoptotic functions.

The cross talk between the BCL-2 family and mitochondrial dynamics is not limited to BAX and BAK. Anti-apoptotic proteins BCL-2, BCL-X$_L$ can associate with MFN2 but not MFN1 in mammalian cells and promote mitochondrial fusion.[116] It is unclear whether BCL-2 and BCL-X$_L$ directly stimulate the GTPase activity of mitofusins. However, the effect of BCL-X$_L$ on mitochondrial morphology is likely more complex as BCL-X$_L$ has also been reported to induce fission.[117] BCL-X$_L$ interacts with Drp1 and stimulates its GTPase activity (Fig. 4).[117] In healthy neurons, DRP1 facilitates disassembly of the mitochondrial tubular network for efficient transport of mitochondria to synapses and plays an important role in establishing and maintaining synapses and dendritic spines.[118,119] BCL-X$_L$ overexpression in rat hippocampal neurons not only increased synaptic transmission, but also induced synapse formation, which may be directly related to its ability to modulate DRP1's function.[117] Recent studies suggested that superimposed on its direct role in mitochondrial fusion/fission, BCL-X$_L$ may also promote mitochondrial biogenesis.[120] As mitochondrial dynamics and biogenesis are tightly integrated with calcium signals,[113] it is likely that the mitochondrial shaping properties of BCL-X$_L$ are closely coupled to its role in synaptic transmission. How the function of mitochondrial fusion/fission factors and select BCL-2 proteins are coordinately cued to regulate mitochondrial dynamics, calcium signals and cellular functions constitutes an exciting area of future experimentation. Furthermore, whether the functional connection between BCL-2 family proteins and mitochondrial dynamics is evolutionarily conserved is an active area of investigation.[116,121,122]

BCL-2 Family and Mitochondrial Fuel Metabolism: A Role for BAD in Glucose Sensing

The link between the BH3-only protein BAD and glucose metabolism was first discovered through proteomic analysis of liver mitochondria that revealed BAD resides in a multi-protein complex together with PKA and PP1, Wiskott-Aldrich family member WAVE-1 and glucokinase (GK, hexokinase IV), the product of *Mody*-2 gene (Maturity Onset Diabetes of the Young type 2).[123] GK is the high *K*m isoform of mammalian hexokinases whose expression is restricted to islet β-cells, hepatocytes and glucose sensing neurons in the hypothalamus.[124] A key component of the mammalian glucose sensing machinery, GK plays important roles in glucose-stimulated insulin secretion in pancreatic β-cells, glucose storage and production in hepatocytes and neuroendocrine regulation of food intake and energy expenditure. The physiologic significance of BAD/GK partnership is underscored by the findings that *Bad* ablation in mice phenocopies multiple parameters associated with GK loss-of-function, including abrogation of glucose-stimulated insulin secretion by β-cells, loss of hepatic glucose sensing and glucose intolerance.[123,125] Initial genetic tests examining the assembly and function of the mitochondria-tethered complex containing BAD and GK provided evidence that BAD is required to nucleate the complex and to support full mitochondria-tethered GK activity. Consistent with this notion, glucose-driven mitochondrial respiration is blunted in *Bad* –/– hepatocytes and primary β-cells.

Mitochondrial localization of hexokinases is thought to couple glycolysis with mitochondrial oxidative phosphorylation in that mitochondria-tethered hexokinases preferentially use the mitochondria-generated ATP to phosphorylate glucose and the ADP produced in the process may in turn activate mitochondrial respiration.[126,127] Of note, recent studies in hepatocytes suggest that mitochondrial localization of GK may represent a functionally distinct pool of this enzyme that is regulated separately from the cytosolic pool likely exerting a dominant effect on oxidative metabolism of glucose and lipogenesis rather than lactate production.[126] On the other hand, mitochondrial association of GK in β-cells may significantly enhance their fuel sensing function by ensuring efficient rise in ATP/ADP ratio, which constitutes the main bioenergetics currency that triggers insulin secretion in response to glucose.[128] The regulatory mechanisms in charge of maintaining the distinct sub-cellular pools of GK are not fully understood. Analogous to other glycolytic enzymes, translocation of GK to distinct sub-cellular compartments may be facilitated through the cytoskeleton. Notably, several glycolytic enzymes, including GK, have been shown to travel on actin filaments within cells.[129-132] The actin cytoskeleton may help focus the glycolytic pathway to site of metabolic demand within the cell. Furthermore, association of these enzymes with actin may help ensure that glycolytic intermediates traverse a shorter distance between subsequent enzymatic steps rather than diffuse in the cytoplasm. In light of these possibilities, the discovery of WAVE-1, which is an AKAP (A Kinase Anchoring Protein) with actin remodeling properties, within the mitochondria-tethered BAD/GK complex is particularly intriguing and warrants investigation of its role in dynamic assembly and recruitment of GK at mitochondria.

The first indication that BAD's role in glucose homeostasis is distinct from its apoptotic function emerged from the observation that *Bad*-null mice and those expressing a nonphosphorylatable mutant of BAD (*Bad*[3SA] knock-in) have common metabolic abnormalities, despite the fact that they represent loss- and gain-of-function models for pro-apoptotic activity of BAD, respectively.[123] While BAD's metabolic function is regulated by its BH3 domain, it does not cosegregate with its ability to engage anti-apoptotic BCL-2 family members.[125] This is consistent with the observation that the metabolic changes associated with *Bad* ablation and transgenic over-expression of *Bcl-x$_L$* in β-cells are not similar.[125,133] Importantly, mutational analysis suggests that phosphorylation of S155 within the BAD BH3 domain [amino acid enumeration based on mouse BAD$_L$ sequence] constitutes a molecular switch between its metabolic and apoptotic functions. When phosphorylated, the BAD BH3 domain engages a metabolic program marked by GK activation, glucose phosphorylation, mitochondrial respiration and ATP production (Fig. 5A). Upon desphosphorylation, the BAD BH3 domain is available to bind anti-apoptotic partners, integrating death signals with the downstream apoptotic machinery at the mitochondrion. Strikingly, the BH3 domain is not only required but sufficient to recapitulate BAD's metabolic activity. BAD BH3 helices that have been chemically reinforced through hydrocarbon stapling (BAD SAHB compounds)[134] and their phosphomimetic derivatives directly bind and activate GK to simulate BAD's effect on glucose metabolism and insulin secretion in β-cells.[125] Interestingly, this property of the BAD BH3 domain is not universally shared with the BH3 domain of other BCL-2 family members despite significant sequence conservation. While the BAD BH3 domain is sufficient to pharmacologically recapitulate BAD's activation of GK, the accessibility of the BH3 domain for binding and activating GK within the context of the entire BAD protein and the BAD/GK mitochondrial signaling

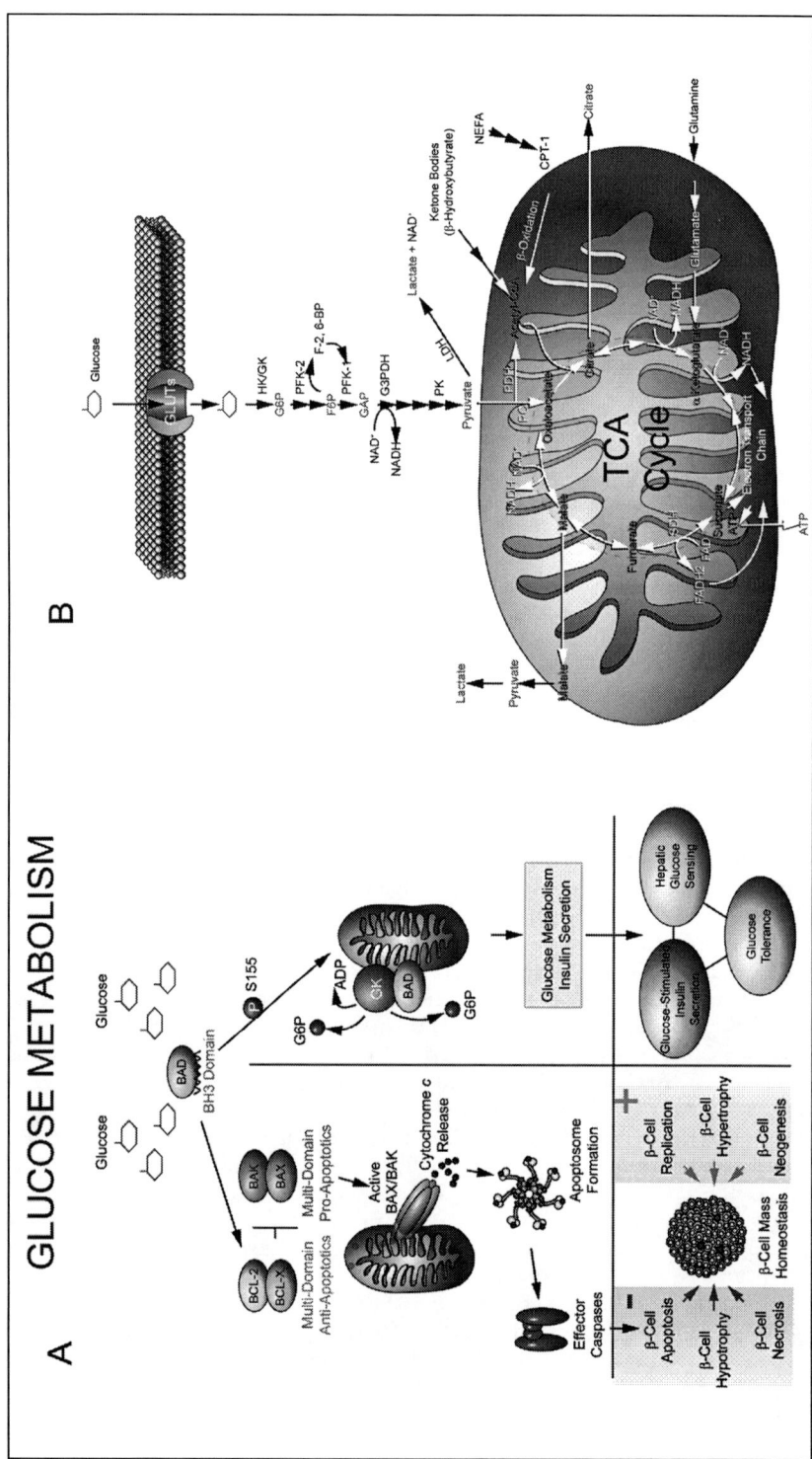

Figure 5. Legend on following page.

Figure 5, viewed on previous page. Intersection of apoptosis and glucose metabolism. (A) Dual role of BAD in glucose metabolism and β-cell survival. Phosphorylation of serine 155 in the BH3 domain instructs BAD to assume a metabolic role by activating GK. The metabolic function of BAD controls insulin secretion, hepatic glucose sensing and overall glucose tolerance. When dephosphorylated, BAD BH3 targets BCL-X$_L$ to induce apoptosis. Apoptosis, together with other negative and positive regulators of β-cell growth, proliferation and survival contribute to physiologic control of β-cell mass homeostasis. (B) Enzymatic steps associated with glycolysis and oxidative metabolism of glucose. Abbreviations: HK, hexokinase, GK, glucokinase, PFK, phosphofructokinase; G3PDH, glyceraldehyde 3-phosphate dehydrogenase; PK, pyruvate kinase, LDH; lactate dehydrogenase; PDH, pyruvate dehydrogenase; PC, pyruvate carboxylase; SDH, succinate dehydrogenase; CPT-1, carnitine palmitoyl transferase-1; G6P, glucose 6-phosphate; F6P, fructose 6-phosphate; GAP; glyceraldehyde phosphate; F-2,6-BP, fructose 2,6-bisphosphate; NEFA, non-esterified fatty acids.

complex is likely more complex. For example, while BAD phosphorylation is not required for the assembly of the entire complex, it is nescessary for activation of GK within the complex. Future studies aimed at in-vitro reconstitution of the complex along with structural information may help shed light on both its stoicheometry and architecture.

In addition to the metabolic benefits of BAD phosphorylation and its pharmacologic mimetics in endocrine tissues such as pancreas and liver, emerging data point to a role for BAD phosphorylation status as an important determinant of β-cell survival, especially in physiologic remodeling of β-cell mass during pancreatic development and under metabolic stress as that associated with nutrient overload and obesity. Importantly, BAD phosphorylation is sensitive to glucose, the nutritional (fed/fasted) state, hormones and growth factors known to enhance β-cell replication and neogenesis [Glucagon-like peptide-1 (GLP-1), IGF and epidermal growth factor (EGF)].[123,125,135,136] Thus, BAD's metabolic and apoptotic activities are likely integrated with nutrient and hormonal regulation of β-cell function and mass, including the developmental remodeling of the endocrine pancreas.[137]

The exact mechanisms by which the nutrient milieu is linked to phosphorylation status of BAD to ultimately instruct the metabolic versus apoptotic roles of this molecule remain to be determined. Many of the BAD kinases identified to date have prominent roles in signaling downstream of nutrient sensing pathways, growth factors and hormone receptors with paramount roles in a variety of tissues, especially endocrine cells [reviewed in ref. 137]. For example, BAD phosphorylation by p70 S6 kinase downstream of the nutrient sensing kinase mTOR (mammalian target of rapamycin) or by p90 ribosomal S6 kinase (RSK) suggests that its posttranslational modification is cued by nutrient availability including amino acid levels.[138,139] Likewise, the ability of PKA to phosphorylate BAD indicates that changes in cAMP levels may influence BAD's function. Moreover, the above-mentioned kinases in charge of BAD phosphorylation have additional phosphorylation targets that include metabolic enzymes and mitochondrial proteins. Thus, BAD may be a component of an intricate signaling circuitry that dynamically reprograms cellular bioenergetics in response to nutritional and hormonal signals.

While GK exerts high control strength over the flux of glucose utilization and the flow of carbons through the glycolytic pathway in hepatocytes and β-cells, low *K*m hexokinases I-III are operative in other tissues. Whether BAD additionally affects the activity of these other hexokinase isoforms remains to be fully explored. Of note, glucose utilization is blunted in mouse embryonic fibroblasts and B-lymphocytes derived from *Bad* –/– mice, (C. Pecqueur and N. Danial, unpublished observations). Recently, another key glycolytic enzyme, phosphofructokinase-1 (PFK-1) (Fig. 5B), was identified as a BAD binding partner in IL-3 dependent hematopoietic FL5.12 cell line.[140] While this association does not depend on BAD's phosphorylation status, stimulation of PFK-1 depends on BAD phosphorylation on Thr201, a target of c-Jun N-terminal protein kinase 1 (JNK-1) downstream of IL-3 signaling pathway. Notably, glycolytic rates are blunted in cells expressing a phosphorylation defective mutant of BAD at this position.[141] Concomitant with a stimulatory effect on glycolysis, Thr201 phosphorylation inhibits BAD's apoptotic activity by interfering with its association with BCL-X$_L$. The nature of BAD phospho-regulatory mechanisms in charge of PFK-1 activation in higher organisms are not fully clear as Thr201 present in rat and mouse BAD (GKGGS**pT**PSQ) is not conserved in human BAD (GRGSSAPSQ). It is possible that alternative mechanisms operate in higher organisms to mediate BAD-regulation of PFK-1 activity. Interestingly, in hepatocytes and pancreatic β-cells, GK binds and activates the bi-functional enzyme 6-phosphofructose-2-kinase/fructose 2,6 bisphosphatase (PFK-2/FBP-2), which generates fructose 2,6,bisphosphate (F2,6BP), the allosteric activator of PFK-1 (Fig. 5B). This ultimately stimulates carbohydrate metabolism.[142] Through its association with GK, BAD may further influence the cellular levels of F2,6BP and subsequent PFK-1 activity. Additional studies are required to examine whether PFK-2/FBP-2 and PFK-1 are components of the core BAD/GK signaling complex. Future dissection of mechanisms by which BAD impacts glycolytic flux in endocrine versus non-endocrine cells should help uncover both common and cell type specific aspects of BAD's metabolic role.

Autophagy

Autophagy is a homeostatic pathway with important roles in organelle turnover, degradation of long-lived proteins, tissue remodeling, genomic stability and cell survival.[143,144] Regulated degradation of proteins and organelles by autophagy is executed by an evolutionarily conserved set of proteins that ultimately regulate the recruitment and packaging of protein/organelle cargo to autophagosoms that will ultimately deliver their contents to lysosomes for degradation by hydrolases. The induction of autophagy is positively regulated by the class III phosphatidyl inositol 3-kinase (PI3K) pathway in association with Beclin 1 (Atg6 in yeast). A lipid kinase signaling complex consisting of Beclin 1, UVRAG, Ambra1 and endophilin B1 (Bif-1) acts as a positive regulator of the class III PI 3-Kinase Vps34, which associates with the myristoylated serine/threonine kinase Vps15 (Fig. 6).[145-147] This lipid signaling complex promotes generation of PI(3)P for vesicle nucleation and formation of the autophagic vacuole.[148] Moreover, the subsequent activation of at least two distinct ubiquitin-like conjugation mechanisms, the Atg12 conjugation system (consisting of Atg5, Atg7, Atg10, Atg12 and Atg16) and the Atg8 conjugation system (consisting of Atg3, Atg4, Atg7 and Atg8), mediate further vesicle expansion and vesicle completion.[149]

Autophagy is tightly coordinated with nutrient and growth factor signaling pathways downstream of PI3K/AKT and mTOR kinases (Fig. 6).[150] The nutrient sensor mTOR is activated in response to elevated ATP, glucose or amino acids and in turn stimulates protein translation and synthesis.[151] In the presence of growth factors and extracellular nutrients, mTOR inhibits autophagy through inactivation of ATG1, an autophagy-related serine/threonine kinase that is important for autophagy induction.[152] Conversely, during nutrient starvation, AMPK inactivates mTOR and releases the "brake" on autophagy (Fig. 6). Upon cellular metabolic decline and nutrient starvation, breakdown of organelles and proteins in autophagosomes produces amino acids and metabolites that can then feed into the mitochondrial TCA cycle to sustain the production of FADH$_2$ and NADH, ensuring that the flow of electrons through the mitochondrial respiratory chain complexes remains uninterrupted. The bioenergetic benefits of autophagy are temporary until the metabolic stress is eliminated (e.g., growth factor or oxygen availability). Inactivation of autophagy during metabolic decline and nutrient stress leads to apoptosis, unless apoptosis is inactivated (e.g., BAX/BAK deficiency), in which case cell death occurs through necrosis.[153] Whether autophagy is primarily a form of cell death or a means of cellular survival is the subject of intense investigation.[154,155] Current findings support the notion that autophagy is primarily a self-limiting survival pathway and a temporary adaptive response during metabolic stress. While excessive autophagy in some experimental settings may be associated with death, the in-vivo significance of autophagy as a death mechanism has been critically questioned.[156]

The discovery of BCL-2 as a binding partner for Beclin 1 suggested that autophagy and survival signaling are highly intertwined homeostatic pathways.[157,158] This interaction is dependent upon the nutritional status of the cells and is directly mediated through a BH3 domain present in Beclin 1.[158-160] Association of Beclin 1 and BCL-2 precludes activation of Vps34 and induction of

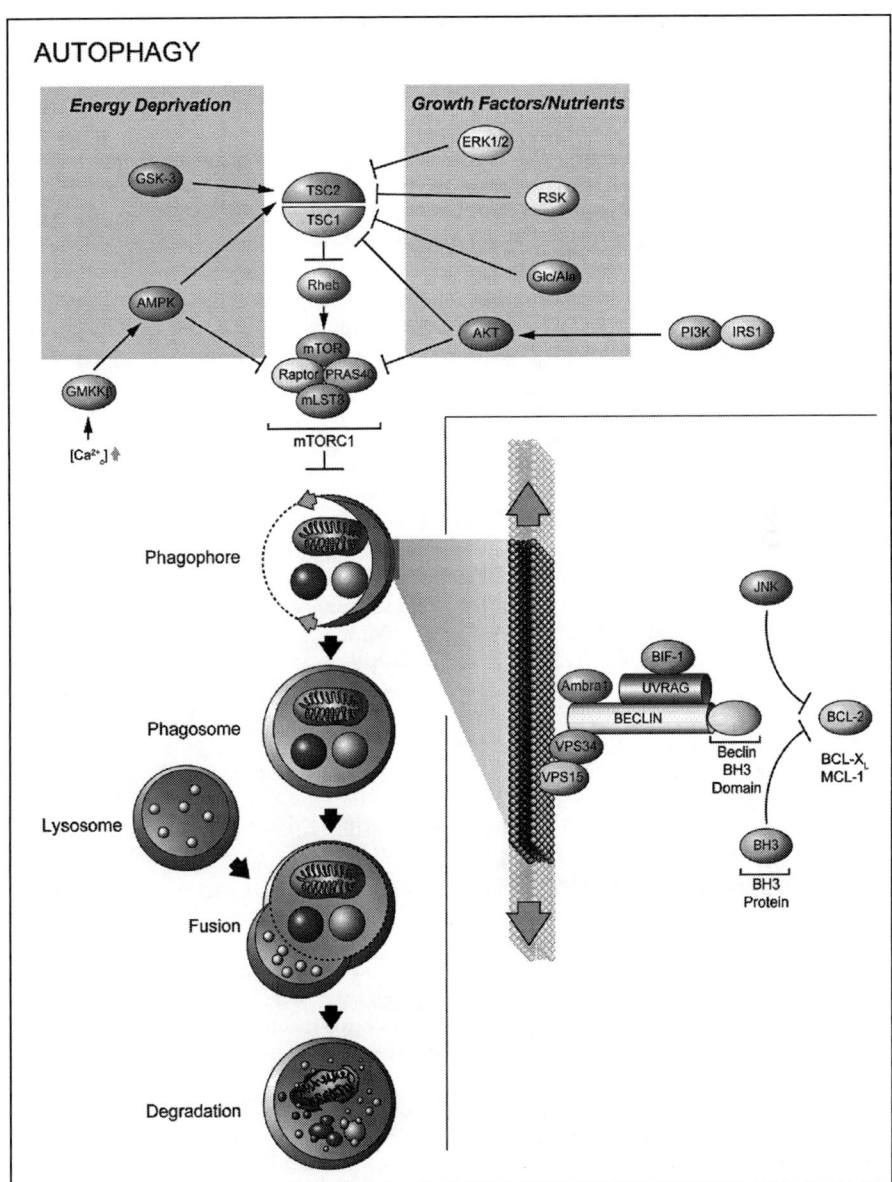

Figure 6. Functional interaction of BCL-2 proteins and Beclin in the control of autophagy.

autophay.[148] Other anti-apoptotic BCL-2 family members, including BCL-X_L, BCL-w and MCL-1 have been shown to inhibit autophagy by direct interaction with Beclin 1.[158,159,161,162] Notably, the capacity of BCL-2 to bind and inhibit Beclin 1 is restricted to the endoplasmic reticulum.

At the cellular level, the interaction of anti-apoptotic BCL-2 family members with Beclin 1 is regulated through multiple mechanisms. For example, BCL-2 phosphorylation at several critical residues in the N-terminal BH4 and BH1 domain[67] inhibits its ability to bind Beclin 1.[163] Notably, JNK1 is rapidly activated upon nutrient deprivation.[163] Consistent with these findings, JNK-dependent activation of autophagy is abolished in cells derived from a nonphosphorylatable *BCL-2* knock-in mouse model.[68,163] Importantly, in JNK1-deficient mouse embryonic fibroblasts, the association between BCL-2 and Beclin 1 cannot be disrupted in response to starvation.[163] Whether JNK1 regulation of BCL-2 and autophagy is restricted to starvation-induced autophagy or is a more general mechanism operative in autophagic turnover of proteins and organelles, tissue remodeling and regulation of genomic integrity warrants investigation. Additional regulatory mechanisms impinging on the interaction of anti-apoptotic BCL-2 proteins and Beclin 1 involve BH3-only proteins such as BNIP3, NIX, BAD, NOXA, PUMA, BIM_{EL} and BIK that may displace Beclin 1 from anti-apoptotic BCL-2 proteins by competing for the same binding site (Fig. 6).[159,164-169] This displacement reaction can also be recapitulated by BH3 mimetic compounds such as ABT737.[159,170] However, the mechanisms by which BH3-only proteins selectively neutralize the ER-resident pool of BCL-2/BCL-X_L to release the inhibition on Beclin 1 are likely complex. Among the BH3-only proteins capable of inducing autophagy, only BIK is localized to the ER.[167,171] Additional studies are required to shed light on the mechanisms utilized by other BH3-only proteins that regulate the interaction of Beclin 1 and BCL2/BCL-X_L at the ER, but are not localized to this organelle. Furthermore, induction of autophagy is likely integrated with ER calcium flux, which is influenced by the anti- and pro-apoptotic BCL-2 family proteins. Beyond starvation, IP3R antagonism or increase in cytosolic calcium by calcium mobilizing agents is associated with increased autophagy due to Ca^{2+}/calmodulin-dependent protein kinase-kinase (CaMKK) inhibition of mTOR that can be countered by BCL-2 and BCL-X_L.[172,173] Of note, Beclin 1 does not appear to impart any influence on ER calcium flux (Fig. 6).[173] It is possible that, in some settings, BCL-2-mediated inhibition of autophagy may be a consequence of its interference with ER calcium flux independent of its direct inhibition of Beclin 1. Likewise, JNK1 phosphorylation of BCL-2 may promote autophagy by promoting calcium release from the ER separate from its interference with BCL-2/Beclin 1 association.

Accumulating evidence also suggests that select neutralization of ER-resident pool of BCL-2/ BCL-X_L by BH3-only proteins or BH3 mimetic compounds is distinct from modulation of apoptosis. For example, over-expression of BAD promoted steady state levels of autophagy in the absence of nutrient starvation.[159] This may be an indication that in healthy cells, in the absence of nutrient decline, BAD may participate in quality control mechanisms that rely on autophagy to rid the cells of damaged organelles, especially dysfunctional mitochondria.[85] Among the BH3-only proteins capable of inducing autophagy, NIX (BNIP3L) appears to have a select role in autophagic elimination of mitochondria (also called mitophagy).[174] Remarkably, NIX loss-of-function is associated with defects in erythroid development that is not related to its pro-apoptotic function.[169,175] During maturation within the bone marrow, reticulocytes normally loose surface area and volume through an autophagy-related process that ensures elimination of ribosomes and organelles.[176-178] Genetic studies have indicated that the removal of mitochondria in this stage of reticulocyte development is NIX dependent.[169]

Protein Quality Control and the Unfolded Protein Response

The endoplasmic reticulum (ER) has evolved to efficiently handle synthesis, folding, processing, assembly and export of newly synthesized proteins marked for secretion or localization to plasma membrane and intracellular organelles. Accumulation of unfolded or misfolded proteins is often associated with aberrations in metabolic, oxidative and bioenergetic balance downstream of physiologic and pathophysiologic stress that trigger an adaptive cellular program known as the unfolded protein response (UPR).[179,180] UPR is primarily a quality control mechanism executed by the ER to ensure proper refolding of misfolded proteins or degradation of those that are not correctly folded or processed. Prolonged UPR transforms into a pro-apoptotic stress response (ER stress) when the homeostatic folding of newly synthesized proteins is not achieved. ER stress can also be associated with genetic/environmental factors and aberrations in Ca^{2+} homeostasis that compromise the proper function of the ER. Cells with high secretory capacity, including islet β-cells and B-lymphocytes, are especially sensitive to UPR as they heavily rely on ER and Ca^{2+} for proper protein folding and secretion of insulin granules and immunoglobulins, respectively.[180,181]

The Sensors and Effectors Downstream of UPR

Three protein sensors, PERK (protein kinase-like ER kinase), ATF-6 (activating transcription factor 6) and IRE1 (Inositol requiring transmembrane kinase/endonuclease 1) are triggered in response to unfolded proteins and activate a homeostatic process that reduces production of new client proteins for the ER folding machinery, helps refold misfolded proteins and degrades protein aggregates.[179,180] The activity of these sensors is normally held dormant due to association with the ER chaperone BiP. During UPR, BiP is bound and sequestered by unfolded proteins leading to derepression of each UPR sensor. PERK is activated by dimerization and autophosphorylation to subsequently phosphorylate the translation initiation factor eIF2α leading to inhibition of general protein translation and selective increase in ATF-4 translation (Fig. 7A). The transcription factor ATF-4 in turn increases expression of select chaperones and antioxidant defense genes. ATF-6 is activated upon translocation to the Golgi and subsequent proteolytic cleavage to a fragment that translocates to the nucleus and binds the UPR response element found in the promoters of target genes. Another UPR sensor, the bifunctional protein kinase IRE1, is activated by dimerization and transphosphorylation, leading to stimulation of its inherent endoribonuclease activity and processing of mRNA encoding the basic leucine zipper transcription factor XBP-1 (X-box binding protein-1). XBP-1 together with ATF-6 regulate transcription of additional genes required for UPR, including chaperones, folding enzymes, protein disulfide isomerase (PDI), ER-associated degradation (ERAD) components and autophagy genes (Fig. 7A). Increased ERAD components and autophagy help clear unfolded protein, protein aggregates and damaged organelles.[182] Remarkably, increased ER biogenesis is also part of the UPR transcriptional program ensuring sufficient ER mass matches this protein quality control response.[183] In addition to alterations in the aforementioned transcriptional programs, select signal transduction pathways and regulated protein-protein interactions contribute to recovery of ER protein overload. For example, release of ER Ca^{2+} subsequent to amassment of unfolded proteins activates autophagy through Ca^{2+}/calmodulin dependent kinase kinase-β[173,184] or protein kinase Cθ(PKC θ) (Fig. 7A).[185] If the integrated outcome of these signaling pathways does not salvage the ER load of unfolded and aggregated proteins, these same UPR sensors can engage the intrinsic pathway of apoptosis.[55,186] P53 and CHOP/GADD153, a transcription factor induced by ATF-4, initiate an ER stress-associated transcription program that is marked by changes in expression levels of several BCL-2 family members,[187-192] death receptors such as FAS and DR5[193] and attenuation of AKT survival pathway through augmented expression of its inhibitor TRB3 (Fig. 7B).[194] Activation of other BH3-only proteins such as BAD and BID at the posttranslational levels can also modulate apoptosis in response to ER stress.[195,196] In addition, recruitment of the adaptor protein TRAF-2 to IRE1[197] may further sensitize cells to ER-stress mediated apoptosis through activation of ER-linked caspases[198] or JNK.[199] JNK-1 phosphorylation of BCL-2 inhibits its survival function (Fig. 7B).[68]

The Outcomes of Signaling Downstream of UPR

While the exact molecular determinants of the switch between UPR as an adaptive/homeostatic process to ER stress as a cellular death response are being actively pursued, emerging evidence from multiple experimental systems indicates that select protein modulators of UPR can be both pro-survival or pro-death depending on the extent of ER damage or the duration of UPR. Recent time course studies evaluating the activation kinetics of the signaling pathway downstream of the individual UPR sensors have been particularly instructive.[200,201] While combined activation of signaling pathways downstream of PERK, ATF-6 and IRE1 is pro-survival during initial phase of UPR, prolonged ER stress is associated with abated IRE1 and ATF-6 activation and persistent PERK activation. The pro-apoptotic consequence of prolonged PERK signaling is consistent with the observation that a chief pro-apoptotic gene downstream of PERK, *Chop*,[202] has a very short half life and may require prolonged stress signaling to reach an apoptotic threshold.[203] Interestingly,

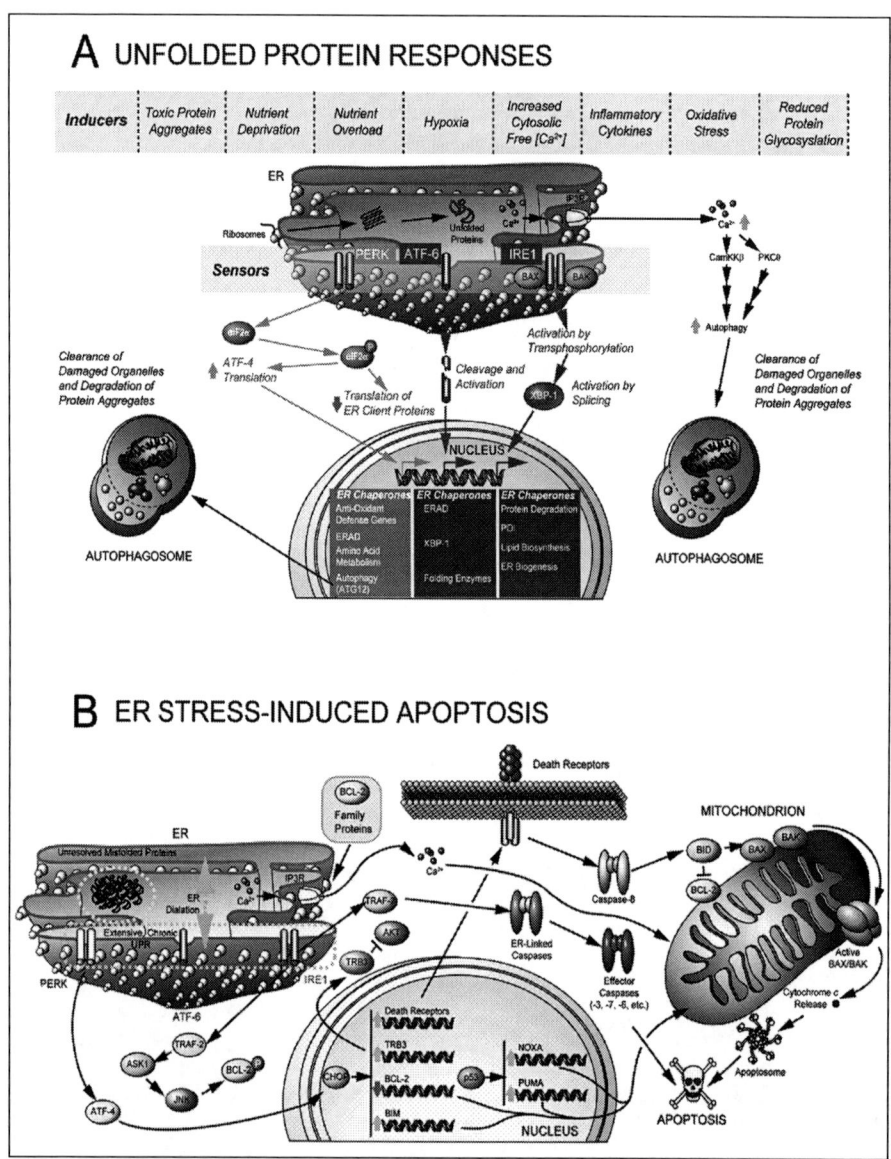

Figure 7. Signaling pathways in response to unfolded proteins. A) The unfolded protein response (UPR) is activated by stimuli such as toxic protein aggregates, nutrient overload, increased cytosolic free [Ca²⁺], inflammatory cytokines and oxidative stress and triggers an adaptive response through sensors that include PERK, ATF-6 and IRE1. Through changes in protein translation and gene expression, UPR leads to refolding of misfolded proteins, degradation of protein aggregates and restoration of ER protein folding homeostasis. B) ER stress ensues upon prolonged UPR and unresolved protein aggregates through the same UPR sensors. *Chop*, an ATF-4 dependent gene downstream of PERK, compromises cell survival by altering the expression levels of several BCL-2 family members and inhibiting the PI3 kinase signaling pathway, while TRAF-2 mediates the apoptotic response downstream of IRE1 through activation of ER-linked caspases and JNK.

forced retention of IRE1 signaling in this setting confers survival benefits,[200] suggesting that IRE1 is sufficient to mitigate the pro-apoptotic effects of persistent PERK activation. How IRE1 may impact cell survival is under active investigation. The discovery of BAX and BAK association with IRE1 and modulation of its downstream effectors, such as XBP-1, suggest a cross talk between BCL-2 family proteins and UPR.[204] The biochemical partnership between BAX/BAK and IRE1 is consistent with the observation that cells doubly deficient in these pro-apoptotic molecules (DKO) display phenotypic similarities with IRE1-deficient cells.[204] Importantly, genetic reconstitution of DKO cells with an ER-targeted BAK restored signaling to XBP-1, suggesting that the role of BAX and BAK in UPR is distinct from their function at mitochondria. How BAX and BAK modulate IRE1 activity/signaling and whether they execute a direct role or an accessory function during each of the adaptive/protective or apoptotic phases of UPR and ER stress remain to be determined. It is possible that BAX and BAK may enforce a conformational change in the kinase domain of IRE1 that triggers its endoribonuclease activity.[205] Indeed, selective activation of IRE1 endoribonuclease activity through chemical genetics promoted cell survival likely through XBP-1.[206] Nevertheless, a role for BAX and BAK in modulation of the unfolded protein response through IRE1 gives rise to the intriguing thesis that they may normally participate in quality control mechanisms regulating the protein folding machinery and cellular protein trafficking separate from their pro-apoptotic function.

Recent studies have also suggested a role for BAK in remodeling the reticular structure of the ER in a manner that is not emulated by BAX but inhibited by BCL-X$_L$, suggesting that the ER localized BAK and BAX may not be functionally redundant.[207] Whether the reticular remodeling property of BAK contributes to ER biogenesis and homeostasis during UPR and is further influenced by its association with IRE1, or whether this ER remodeling is solely integral to BAK's apoptotic function is unclear. As various members of the BCL-2 family are known to localize to different ER-resident complexes, including IP3R[41] and Beclin 1[158] the availability of BAX and BAK to biochemically and functionally interact with IRE1 may be under complex and dynamic regulation by other ER-resident BCL-2 proteins.

Physiologic Roles for BCL-2 Family Proteins in Cell Cycle Checkpoints, Differentiation and DNA Damage Response

A growing body of evidence suggests an intimate connection between cell cycle progression and the cellular life and death decisions. The role of several cell cycle regulators such as p53, MYC and the E2F family of transcription factors in apoptosis has been well established.[208-210] Similarly, cell cycle regulatory effects for multiple components of the cell death pathway, including the inhibitor of apoptosis protein (IAP) Survivin,[211] caspases[212] and select BCL-2 family members[213] have been recognized. Cell cycle progression is a carefully coordinated process that receives input from multiple genome surveillance mechanisms, including DNA damage response, DNA repair and chromatin remodeling processes to preserve genomic stability. Whether a cell commits to reenter the cell cycle, undergo senescence or execute apoptosis depends on the integrated outcome of these processes. The following sections provide a brief overview of recent findings from multiple lines of investigation implicating select BCL-2 family proteins in cell cycle progression, differentiation and DNA damage checkpoints.

BCL-2 Family and Cell Cycle Checkpoints

Anti-apoptotic BCL-2, BCL-X$_L$ and MCL-1 delay cell cycle entry through distinct mechanisms.[214-219] BCL-2 and BCL-X$_L$ prolong G0 to delay S phase entry by sustaining the levels of the cyclin CDK inhibitor p27 and by increasing the expression of p130, which binds E2F4 to repress transcription of target genes required for cell cycle progression (Fig. 8A).[217,220-223] Similar to *BCL-2* transgenic T-cells, *Bax$^{-/-}$Bak$^{-/-}$* thymocytes show a proliferative defect upon T-cell receptor activation.[224] Curiously, the effect of BCL-X$_L$ on p27 stabilization requires intact BAX and BAK and may be further modulated by BH3-only proteins such as BIM.[225] Loss- and gain-of-function studies, acute knock-down experiments using small interfering RNA (siRNA) and genetic reconstitution

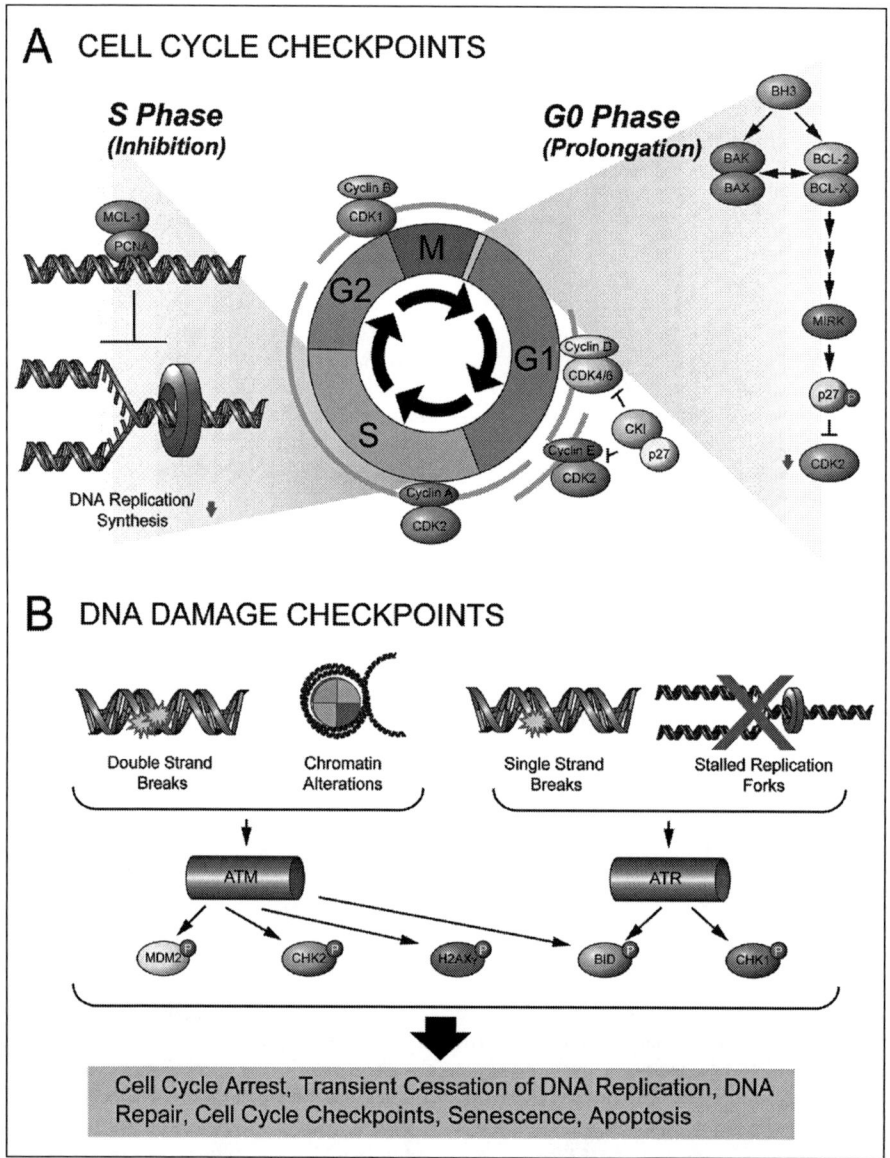

Figure 8. BCL-2 proteins in cell cycle (A) and DNA damage (B) checkpoints.

assays suggest that BAX and BAK normally inhibit phosphorylation of p27 by the arginine-directed serine/threonine kinase Mirk.[225,226] This post translational modification is tightly regulated during G0 and leads to stabilization of the p27 protein levels (Fig. 8A).[227] How Mirk activity is selectively activated during G0 by BAX and BAK is unknown. In light of these observations, it is possible that the sustained p27 levels associated with BCL-2 and BCL-X_L gain-of-function may reflect their counter-regulatory effect on BAX and BAK, which can be further modulated by BH3-only proteins. The BH3-only protein BAD counters G0/G1 arrest induced by serum withdrawal or over-expression of anti-apoptotic BCL-2 and BCL-X_L and aberrantly drives cells through the S phase.[228-230] BAD's effect on cell cycle depends on its BH3 domain, requires its ability to engage BCL-2/BCL-X$_L$ and is associated with persistent CDK2 activity.[228]

Although the cell death regulatory function of BCL-2 and BCL-X$_L$ and their antiproliferative effects appear to share common biochemical properties, several findings suggest that their cell cycle function is not a mere byproduct of their pro-survival activity. For example, treatment of cells with a pan-caspase inhibitor or a dominant negative caspase-9 mutant in the absence of growth factors did not recapitulate the effect of BCL-2 or BCL-X$_L$ in sustaining p27 levels or prolonging G0.[217,225] However, analysis of defined mutations to genetically and biochemically segregate the cell cycle effect and the pro-survival role of BCL-2 and BCL-X$_L$ has not been conclusive. For example, mutations within the unstructured loop between the BH3 and BH4 domains of BCL-2[231] or the Y28 residue within the BH4 domain (Y28A, Y28F and Y28S) were shown to retain pro-survival function while incapable of retarding cell cycle progression in some studies[223,232] but not others.[229,233]

Among potential mechanisms underlying the cell cycle effect of BCL-2 family proteins, a role for alterations in cellular bioenergetics can be entertained. For example, the capacity of BCL-X$_L$ to regulate the permeability of mitochondria to metabolites (see section I and ref. 13), as well as the intracellular Ca^{2+} stores (see section II and ref. 234) and redox state[235,236] can ultimately impact cellular ATP, ROS and Ca^{2+} flux that could ultimately modulate cell cycle components such as p27 and CDK-2. This scenario is in line with recent studies demonstrating that cellular energy balance, specifically the efficiency of mitochondrial electron transport, impacts cell cycle progression by regulating cyclin E-CDK2 activity.[237,238] Notably, genetic dissection of the cross talk between mitochondria and cell cycle machinery in *Drosophila* has revealed that two indicators of mitochondrial electron transport efficiency, ATP and ROS levels, influence cyclin E-CDK2 activity by controlling the activation status of AMPK and JNK, respectively.[238] AMPK is a cellular energy sensor activated by rising AMP levels, which mirrors the drop in ATP/ADP ratio. Through its downstream substrates, AMPK stimulates catabolic pathways and inactivates ATP consuming anabolic pathways such as protein synthesis and cell cycle progression.[239] AMPK phosphorylation of p53 leads to its cell cycle inhibitory function, including down-regulation of cyclin E levels.[240,241] AMPK phosphorylation also increases the protein stability of p27.[242] In the context of suboptimal mitochondrial function, p27 levels are further increased in response to moderate elevation of ROS and subsequent JNK activation.[238] It is important to note that increased levels of ROS is not always associated with cell cycle arrest; indeed, select ROS are required for proliferation in certain cell types. For example, hydrogen peroxide is necessary for phosphorylation and activation of extracellular signal regulated kinase (ERK) upon T-cell receptor stimulation,[243] which can be further influenced by BCL-2 family members, such as BAX and BAK.[224]

Differentiation

The antiproliferative properties of BCL-2 and BCL-X$_L$ may directly instruct differentiation and cellular fate decisions. For example, stable expression of BCL-2 in the FDCP-Mix hematopoietic progenitor cells is associated with differentiation into myeloid lineage in the absence of IL-3.[244] Interestingly, stable expression of BCL-X$_L$, which similarly promotes survival and retards cell cycle progression, led to acquisition of erythroid cell fate.[244] These findings indicate that BCL-2 and BCL-X$_L$ initiate a specific cell fate program rather than playing a more permissive role due to their anti-apoptotic effects. Remarkably, grafting the BH4 sequence of BCL-X$_L$ onto BCL-2 was sufficient to activate the erythroid differentiation program, suggesting a role for the BH4 domain in lineage commitment. The lineage decision driven by BCL-2 appears to be regulated by Raf-1, which upon binding to the BCL-2 BH4 domain, is subjected to ubiquitin-mediated degradation.[244] Consistent with this, expression of Raf-1 or inhibition of its degradation is sufficient to direct differentiation of the BCL-2 expressing FDCP-Mix cells towards an erythroid cell fate in the absence of IL-3. Curiously, the participation of Raf-1 in this pathway is independent of its kinase activity. The requirement of other BH4-interacting proteins such as the phosphatase calcineurin,[245] the mitochondrial chaperone FKBP38[246] and Ras[247] in this myeloid differentiation program awaits loss of function studies. While Raf-1 appears to be a possible effector molecule through which BCL-2 drives myeloid differentiation, molecules mediating the differentiation function of BCL-X$_L$ have not been fully explored. For example, although Raf-1 does not undergo degradation in progenitors expressing BCL-X$_L$, its absolute requirement in erythroid lineage commitment has not been formally examined.

Regulation of cellular differentiation by BCL-2 is not restricted to the hematopoietic cells. Loss- and gain-of-function studies have supported a role for BCL-2 in neuronal differentiation, sprouting and axonal regeneration.[248-251] A significant increase in regeneration of retinal axons was observed in neuronal specific *BCL-2* transgene upon optic nerve injury, which was not accompanied by changes in neuronal survival and could not be simulated by caspase inhibition or treatment with neurotrophins.[248] In addition, BCL-2 expression in CNS axons leads to greater influx of Ca^{2+} supporting axonal regeneration and growth through cAMP-mediated activation of cAMP response element binding protein (CREB) and ERK in a manner independent of its survival function.[252] The BCL-2 interacting protein Bag-1, which has also been implicated in neuronal differentiation and axonal regeneration,[253,254] may be a potential mediator of BCL-2's effect on axonal regeneration, however this possibility has not been formally examined.

DNA Replication and Damage Checkpoints

While the antiproliferative effect of BCL-2 and BCL-X$_L$ impinge on G0, the cell cycle function of MCL-1 appears to be restricted to the S phase (Fig. 8A).[215] Inhibition of DNA synthesis by MCL-1 is mediated, at least in part, through its association with proliferating cell nuclear antigen (PCNA),[215] which serves as a cofactor for DNA polymerase δ to ensure processivity of its polymerase activity. In addition, PCNA provides a molecular platform to recruit proteins involved in DNA repair and chromatin assembly, ensuring that these processes are coordinated with DNA replication.[255] The interaction between MCL-1 and PCNA is mediated through a consensus PCNA binding motif in the amino acid sequence between the BH1 and BH3 domains of MCL-1 and the C-terminal half of PCNA.[215] Notably, the anti-apoptotic and cell cycle functions of MCL-1 appear to be mediated by distinct regions as mutants deficient in PCNA binding and S phase inhibition did not show any alterations in their anti-apoptotic potency.[215] The molecular mechanisms by which MCL-1 may interfere with DNA replication through its association with PCNA have not been explored. MCL-1 may influence PCNA function or its dynamic loading and unloading unto DNA by competing with other PCNA interacting molecules or by influencing its post translational modifications known to influence PCNA's function.[255] Furthermore, a direct modulation of DNA polymerase by MCL-1 has not been ruled out.

Errors and lesions accumulated during DNA replication or DNA damage during S phase pose a significant risk to genomic integrity and are tightly monitored and resolved by an intra-S checkpoint program composed of DNA damage sensors, mediators and effectors.[256,257] Two DNA damage kinases, Ataxia-Telangiectasia Mutated (ATM) and ATM-and Rad 3-related (ATR) kinases, are recruited to double strand breaks and single strand breaks or stalled replication forks, respectively (Fig. 8B).[256] ATM and ATR phosphorylate a series of downstream effector molecules, leading to cell cycle arrest and DNA repair. Depending on the extent of DNA damage or the outcome of repair and checkpoint recovery, the cells may re-enter cell cycle, permanently withdraw to become senescent or undergo apoptosis. While the interconnection between DNA repair machinery and cell cycle regulators has been more extensively studied, the cross talk between DNA repair and the apoptotic machinery is just beginning to be unraveled.

The BH3-only pro-apoptotic protein BID was recently found to activate an intra-S checkpoint in response to inhibitors of DNA replication or select genotoxic agents invoking replication stress.[258-260] In response to these agents, BID-deficient myeloid cells and T-lymphocytes do not arrest in S phase, continue to replicate their DNA and are significantly more sensitive to apoptosis compared to WT cells.[260] A defect in execution of an S phase checkpoint and recovery through DNA repair is also consistent with the observed accumulation of chromosomal aberrations in these cells. A role for BID in DNA damage checkpoint is also consistent with its translocation to the nucleus and association with chromatin upon DNA damage.[258-260] The function of BID in DNA damage checkpoint is independent of its BH3 domain and is activated primarily by ATM and ATR DNA damage kinases, which phosphorylate BID on a serine residues N terminal to its BH3 domain (Fig. 8B).[258,260] By serving as a downstream substrate of DNA damage surveillance and repair pathways, BID is embedded in a DNA damage checkpoint that likely operates parallel to its pro-apoptotic function, instructing cells to undergo DNA repair or apoptosis in response to DNA damage. As several DNA damage checkpoint proteins appear to play a role during normal progression of S phase likely by influencing the progression of replication forks or the firing time of replication origins,[261] investigating the significance of BID in normal S phase progression would provide important insights. This is especially relevant as a portion of BID in healthy cells is localized to the nucleus.[258] Furthermore, given the localization of BID to chromatin upon DNA damage, exploring its potential direct role in DNA damage surveillance and repair is highly warranted. Lastly, whether and how the functions of nuclear and mitochondrial pools of BID with respective functions in DNA damage checkpoint and apoptosis are coordinated in a spatiotemporal manner would be of great interest.

Conclusion

A growing body of evidence implicates BCL-2 proteins as active components of select physiologic pathways carrying "day jobs" beyond regulation of apoptosis. As such, targeting BCL-2 proteins for therapeutic purpose may additionally require our careful assessment of these alternative functions. We anticipate that the expanding functional interaction network of BCL-2 proteins will include additional noncell death related proteins, including alternative BH3-ligands. Within this exciting area of investigation, there lies enormous potential in discovering molecular mechanisms that selectively activate/deactivate these apoptotic and non-apoptotic functions of BCL-2 proteins. Such endeavors may help uncover how cellular life and death decisions are integrated with other physiologic pathways to achieve a singular biologic mission, namely homeostasis.

Acknowledgements

The authors would like to thank Eric Smith for creation of figures and editorial assistance.

References

1. Chipuk JE, Green DR. How do BCL-2 proteins induce mitochondrial outer membrane permeabilization? Trends Cell Biol 2008; 18(4):157-164.
2. Danial NN, Korsmeyer SJ. Cell death: critical control points. Cell 2004; 116(2):205-219.
3. Youle RJ, Strasser A. The BCL-2 protein family: opposing activities that mediate cell death. Nat Rev Mol Cell Biol 2008; 9(1):47-59.
4. Leber B, Lin J, Andrews DW. Embedded together: the life and death consequences of interaction of the BCL-2 family with membranes. Apoptosis 2007; 12(5):897-911.
5. Lindsten T, Ross AJ, King A et al. The combined functions of proapoptotic BCL-2 family members bak and bax are essential for normal development of multiple tissues. Mol Cell 2000; 6(6):1389-1399.
6. Wei MC, Zong WX, Cheng EH et al. Proapoptotic BAX and BAK: a requisite gateway to mitochondrial dysfunction and death. Science 2001; 292(5517):727-730.
7. Cheng WC, Berman SB, Ivanovska I et al. Mitochondrial factors with dual roles in death and survival. Oncogene 2006; 25(34):4697-4705.
8. Colombini M. VDAC: the channel at the interface between mitochondria and the cytosol. Mol Cell Biochem 2004; 256-257(1-2):107-115.
9. Rostovtseva TK, Tan W, Colombini M. On the role of VDAC in apoptosis: fact and fiction. J Bioenerg Biomembr 2005; 37(3):129-142.
10. Baines CP, Kaiser RA, Sheiko T et al. Voltage-dependent anion channels are dispensable for mitochondrial-dependent cell death. Nat Cell Biol 2007; 9(5):550-555.
11. Malia TJ, Wagner G. NMR structural investigation of the mitochondrial outer membrane protein VDAC and its interaction with antiapoptotic Bcl-xL. Biochemistry 2007; 46(2):514-525.
12. Shimizu S, Narita M, Tsujimoto Y. BCL-2 family proteins regulate the release of apoptogenic cytochrome c by the mitochondrial channel VDAC. Nature 1999; 399(6735):483-487.

13. Vander Heiden MG, Li XX, Gottleib E et al. Bcl-xL promotes the open configuration of the voltage-dependent anion channel and metabolite passage through the outer mitochondrial membrane. J Biol Chem 2001; 276(22):19414-19419.

14. Lee AC, Zizi M, Colombini M. Beta-NADH decreases the permeability of the mitochondrial outer membrane to ADP by a factor of 6. J Biol Chem 1994; 269(49):30974-30980.

15. Hiller S, Garces RG, Malia TJ et al. Solution structure of the integral human membrane protein VDAC-1 in detergent micelles. Science 2008; 321(5893):1206-1210.

16. Schwarzer C, Barnikol-Watanabe S, Thinnes FP et al. Voltage-dependent anion-selective channel (VDAC) interacts with the dynein light chain Tctex1 and the heat-shock protein PBP74. Int J Biochem Cell Biol 2002; 34(9):1059-1070.

17. Carre M, Andre N, Carles G et al. Tubulin is an inherent component of mitochondrial membranes that interacts with the voltage-dependent anion channel. J Biol Chem 2002; 277(37):33664-33669.

18. Rostovtseva TK, Sheldon KL, Hassanzadeh E et al. Tubulin binding blocks mitochondrial voltage-dependent anion channel and regulates respiration. Proc Natl Acad Sci USA 2008; 105(48):18746-18751.

19. Xu X, Forbes JG, Colombini M. Actin modulates the gating of Neurospora crassa VDAC. J Membr Biol 2001; 180(1):73-81.

20. Pastorino JG, Hoek JB. Regulation of hexokinase binding to VDAC. J Bioenerg Biomembr 2008; 40(3):171-182.

21. Rostovtseva TK, Kazemi N, Weinrich M et al. Voltage gating of VDAC is regulated by nonlamellar lipids of mitochondrial membranes. J Biol Chem 2006; 281(49):37496-37506.

22. Baines CP, Song CX, Zheng YT et al. Protein kinase Cepsilon interacts with and inhibits the permeability transition pore in cardiac mitochondria. Circ Res 2003; 92(8):873-880.

23. Bera AK, Ghosh S. Dual mode of gating of voltage-dependent anion channel as revealed by phosphorylation. J Struct Biol 2001; 135(1):67-72.

24. Pastorino JG, Hoek JB, Shulga N. Activation of glycogen synthase kinase 3beta disrupts the binding of hexokinase II to mitochondria by phosphorylating voltage-dependent anion channel and potentiates chemotherapy-induced cytotoxicity. Cancer Res 2005; 65(22):10545-10554.

25. Le Mellay V, Troppmair J, Benz R et al. Negative regulation of mitochondrial VDAC channels by C-Raf kinase. BMC Cell Biol 2002; 3:14.

26. Roy SS, Madesh M, Davies E et al. Bad targets the permeability transition pore independent of Bax or Bak to switch between Ca^{2+}-dependent cell survival and death. Mol Cell 2009; 33(3):377-388.

27. Mattson MP, Gleichmann M, Cheng A. Mitochondria in neuroplasticity and neurological disorders. Neuron 2008; 60(5):748-766.

28. Jonas EA, Buchanan J, Kaczmarek LK. Prolonged activation of mitochondrial conductances during synaptic transmission. Science 1999; 286(5443):1347-1350.

29. Jonas EA, Hoit D, Hickman JA et al. Modulation of synaptic transmission by the BCL-2 family protein BCL-xL. J Neurosci 2003; 23(23):8423-8431.

30. Hickman JA, Hardwick JM, Kaczmarek LK et al. Bcl-xL inhibitor ABT-737 reveals a dual role for Bcl-xL in synaptic transmission. J Neurophysiol 2008; 99(3):1515-1522.

31. Duchen MR. Mitochondria and calcium: from cell signalling to cell death. J Physiol 2000; 529 Pt 1:57-68.

32. Graier WF, Frieden M, Malli R. Mitochondria and Ca(2+) signaling: old guests, new functions. Pflugers Arch 2007; 455(3):375-396.

33. Csordas G, Thomas AP, Hajnoczky G. Quasi-synaptic calcium signal transmission between endoplasmic reticulum and mitochondria. EMBO J 1999; 18(1):96-108.

34. Rizzuto R, Brini M, Murgia M et al. Microdomains with high Ca^{2+} close to IP3-sensitive channels that are sensed by neighboring mitochondria. Science 1993; 262(5134):744-747.

35. Rizzuto R, Pinton P, Carrington W et al. Close contacts with the endoplasmic reticulum as determinants of mitochondrial Ca^{2+} responses. Science 1998; 280(5370):1763-1766.

36. Csordas G, Renken C, Varnai P et al. Structural and functional features and significance of the physical linkage between ER and mitochondria. J Cell Biol 2006; 174(7):915-921.

37. de Brito OM, Scorrano L. Mitofusin 2 tethers endoplasmic reticulum to mitochondria. Nature 2008; 456(7222):605-610.

38. Simmen T, Aslan JE, Blagoveshchenskaya AD et al. PACS-2 controls endoplasmic reticulum-mitochondria communication and Bid-mediated apoptosis. EMBO J 2005; 24(4):717-729.

39. Berridge MJ, Bootman MD, Roderick HL. Calcium signalling: dynamics, homeostasis and remodelling. Nat Rev Mol Cell Biol 2003; 4(7):517-529.

40. Orrenius S, Zhivotovsky B, Nicotera P. Regulation of cell death: the calcium-apoptosis link. Nat Rev Mol Cell Biol 2003; 4(7):552-565.

41. Rong Y, Distelhorst CW. BCL-2 protein family members: versatile regulators of calcium signaling in cell survival and apoptosis. Annu Rev Physiol 2008; 70:73-91.
42. Germain M, Mathai JP, McBride HM et al. Endoplasmic reticulum BIK initiates DRP1-regulated remodelling of mitochondrial cristae during apoptosis. EMBO 2005; 24(8):1546-1556.
43. Mathai JP, Germain M, Shore GC. BH3-only BIK regulates BAX, BAK-dependent release of Ca^{2+} from endoplasmic reticulum stores and mitochondrial apoptosis during stress-induced cell death. J Biol Chem 2005; 280(25):23829-23836.
44. Morishima N, Nakanishi K, Tsuchiya K et al. Translocation of Bim to the endoplasmic reticulum (ER) mediates ER stress signaling for activation of caspase-12 during ER stress-induced apoptosis. J Biol Chem 2004; 279(48):50375-50381.
45. Scorrano L, Oakes SA, Opferman JT et al. BAX and BAK regulation of endoplasmic reticulum Ca^{2+}: a control point for apoptosis. Science 2003; 300(5616):135-139.
46. Zong WX, Li C, Hatzivassiliou G et al. Bax and Bak can localize to the endoplasmic reticulum to initiate apoptosis. J Cell Biol 2003; 162(1):59-69.
47. Baffy G, Miyashita T, Williamson JR et al. Apoptosis induced by withdrawal of interleukin-3 (IL-3) from an IL-3-dependent hematopoietic cell line is associated with repartitioning of intracellular calcium and is blocked by enforced BCL-2 oncoprotein production. J Biol Chem 1993; 268(9):6511-6519.
48. Carvalho AC, Sharpe J, Rosenstock TR et al. Bax affects intracellular Ca^{2+} stores and induces Ca^{2+} wave propagation. Cell Death Differ 2004; 11(12):1265-1276.
49. Csordas G, Madesh M, Antonsson B et al. tcBid promotes Ca(2+) signal propagation to the mitochondria: control of Ca(2+) permeation through the outer mitochondrial membrane. EMBO J 2002; 21(9):2198-2206.
50. Lam M, Dubyak G, Chen L et al. Evidence that BCL-2 represses apoptosis by regulating endoplasmic reticulum-associated Ca^{2+} fluxes. Proc Natl Acad Sci USA 1994; 91(14):6569-6573.
51. Nutt LK, Pataer A, Pahler J et al. Bax and Bak promote apoptosis by modulating endoplasmic reticular and mitochondrial Ca^{2+} stores. J Biol Chem 2002; 277(11):9219-9225.
52. Pinton P, Ferrari D, Rapizzi E et al. The Ca^{2+} concentration of the endoplasmic reticulum is a key determinant of ceramide-induced apoptosis: significance for the molecular mechanism of BCL-2 action. EMBO J 2001; 20(11):2690-2701.
53. Wang NS, Unkila MT, Reineks EZ et al. Transient expression of wild-type or mitochondrially targeted BCL-2 induces apoptosis, whereas transient expression of endoplasmic reticulum-targeted BCL-2 is protective against Bax-induced cell death. J Biol Chem 2001; 276(47):44117-44128.
54. White C, Li C, Yang J et al. The endoplasmic reticulum gateway to apoptosis by Bcl-X(L) modulation of the InsP3R. Nat Cell Biol 2005; 7(10):1021-1028.
55. Heath-Engel HM, Chang NC, Shore GC. The endoplasmic reticulum in apoptosis and autophagy: role of the BCL-2 protein family. Oncogene 2008; 27(50):6419-6433.
56. Pinton P, Giorgi C, Siviero R et al. Calcium and apoptosis: ER-mitochondria Ca^{2+} transfer in the control of apoptosis. Oncogene 2008; 27(50):6407-6418.
57. Chami M, Prandini A, Campanella M et al. BCL-2 and Bax exert opposing effects on Ca^{2+} signaling, which do not depend on their putative pore-forming region. J Biol Chem 2004; 279(52):54581-54589.
58. Chen R, Valencia I, Zhong F et al. BCL-2 functionally interacts with inositol 1,4,5-trisphosphate receptors to regulate calcium release from the ER in response to inositol 1,4,5-trisphosphate. J Cell Biol 2004; 166(2):193-203.
59. Oakes SA, Scorrano L, Opferman JT et al. Proapoptotic BAX and BAK regulate the type 1 inositol trisphosphate receptor and calcium leak from the endoplasmic reticulum. Proc Natl Acad Sci USA 2005; 102(1):105-110.
60. Dremina ES, Sharov VS, Kumar K et al. Anti-apoptotic protein BCL-2 interacts with and destabilizes the sarcoplasmic/endoplasmic reticulum Ca^{2+}-ATPase (SERCA). Biochem J 2004; 383(Pt 2):361-370.
61. Dremina ES, Sharov VS, Schoneich C. Displacement of SERCA from SR lipid caveolae-related domains by BCL-2: a possible mechanism for SERCA inactivation. Biochemistry 2006; 45(1):175-184.
62. Kuo TH, Kim HR, Zhu L et al. Modulation of endoplasmic reticulum calcium pump by BCL-2. Oncogene 1998; 17(15):1903-1910.
63. Distelhorst CW, Shore GC. BCL-2 and calcium: controversy beneath the surface. Oncogene 2004; 23(16):2875-2880.
64. Palmer AE, Jin C, Reed JC et al. BCL-2-mediated alterations in endoplasmic reticulum Ca^{2+} analyzed with an improved genetically encoded fluorescent sensor. Proc Natl Acad Sci USA 2004; 101(50):17404-17409.
65. Pinton P, Ferrari D, Magalhaes P et al. Reduced loading of intracellular Ca(2+) stores and downregulation of capacitative Ca(2+) influx in BCL-2-overexpressing cells. J Cell Biol 2000; 148(5):857-862.
66. Erin N, Bronson SK, Billingsley ML. Calcium-dependent interaction of calcineurin with BCL-2 in neuronal tissue. Neuroscience 2003; 117(3):541-555.

67. Yamamoto K, Ichijo H, Korsmeyer SJ. BCL-2 Is Phosphorylated and Inactivated by an ASK1/Jun N-Terminal Protein Kinase Pathway Normally Activated at G(2)/M. Mol Cell Biol 1999; 19(12):8469-8478.
68. Bassik MC, Scorrano L, Oakes SA et al. Phosphorylation of BCL-2 regulates ER Ca²⁺ homeostasis and apoptosis. EMBO J 2004; 23(5):1207-1216.
69. Booth C, Koch GL. Perturbation of cellular calcium induces secretion of luminal ER proteins. Cell 1989; 59(4):729-737.
70. Zhong F, Davis MC, McColl KS et al. BCL-2 differentially regulates Ca²⁺ signals according to the strength of T-cell receptor activation. J Cell Biol 2006; 172(1):127-137.
71. Hajnoczky G, Robb-Gaspers LD, Seitz MB et al. Decoding of cytosolic calcium oscillations in the mitochondria. Cell 1995; 82(3):415-424.
72. Lasorsa FM, Pinton P, Palmieri L et al. Recombinant expression of the Ca(2+)-sensitive aspartate/glutamate carrier increases mitochondrial ATP production in agonist-stimulated Chinese hamster ovary cells. J Biol Chem 2003; 278(40):38686-38692.
73. Denton RM, McCormack JG, Edgell NJ. Role of calcium ions in the regulation of intramitochondrial metabolism. Effects of Na+, Mg2+ and ruthenium red on the Ca²⁺-stimulated oxidation of oxoglutarate and on pyruvate dehydrogenase activity in intact rat heart mitochondria. Biochem J 1980; 190(1):107-117.
74. Robb-Gaspers LD, Burnett P, Rutter GA et al. Integrating cytosolic calcium signals into mitochondrial metabolic responses. EMBO 1998; 17(17):4987-5000.
75. Territo PR, Mootha VK, French SA et al. Ca(2+) activation of heart mitochondrial oxidative phosphorylation: role of the F(0)/F(1)-ATPase. Am J Physiol Cell Physiol 2000; 278(2):C423-435.
76. Jouaville LS, Pinton P, Bastianutto C et al. Regulation of mitochondrial ATP synthesis by calcium: evidence for a long-term metabolic priming. Proc Natl Acad Sci USA 1999; 96(24):13807-13812.
77. Szalai G, Krishnamurthy R, Hajnoczky G. Apoptosis driven by IP(3)-linked mitochondrial calcium signals. EMBO J 1999; 18(22):6349-6361.
78. Pacher P, Hajnoczky G. Propagation of the apoptotic signal by mitochondrial waves. EMBO J 2001; 20(15):4107-4121.
79. Berridge MJ, Lipp P, Bootman MD. The versatility and universality of calcium signalling. Nat Rev Mol Cell Biol 2000; 1(1):11-21.
80. Frey TG, Mannella CA. The internal structure of mitochondria. Trends Biochem Sci 2000; 25(7):319-324.
81. Mannella CA. Structure and dynamics of the mitochondrial inner membrane cristae. Biochim Biophys Acta 2006; 1763(5-6):542-548.
82. Hackenbrock CR. Ultrastructural bases for metabolically linked mechanical activity in mitochondria. I. Reversible ultrastructural changes with change in metabolic steady state in isolated liver mitochondria. J Cell Biol 1966; 30(2):269-297.
83. Chen H, Chan DC. Mitochondrial dynamics in mammals. Curr Top Dev Biol 2004; 59:119-144.
84. Shaw JM, Nunnari J. Mitochondrial dynamics and division in budding yeast. Trends Cell Biol 2002; 12(4):178-184.
85. Twig G, Elorza A, Molina AJ et al. Fission and selective fusion govern mitochondrial segregation and elimination by autophagy. EMBO 2008; 27(2):433-446.
86. Bleazard W, McCaffery JM, King EJ et al. The dynamin-related GTPase Dnm1 regulates mitochondrial fission in yeast. Nat Cell Biol 1999; 1(5):298-304.
87. James DI, Parone PA, Mattenberger Y et al. hFis1, a novel component of the mammalian mitochondrial fission machinery. J Biol Chem 2003; 278(38):36373-36379.
88. Labrousse AM, Zappaterra MD, Rube DA et al. C. elegans dynamin-related protein DRP-1 controls severing of the mitochondrial outer membrane. Mol Cell 1999; 4(5):815-826.
89. Mozdy AD, McCaffery JM, Shaw JM. Dnm1p GTPase-mediated mitochondrial fission is a multi-step process requiring the novel integral membrane component Fis1p. J Cell Biol 2000; 151(2):367-380.
90. Smirnova E, Griparic L, Shurland DL et al. Dynamin-related protein Drp1 is required for mitochondrial division in mammalian cells. Mol Biol Cell 2001; 12(8):2245-2256.
91. Stojanovski D, Koutsopoulos OS, Okamoto K et al. Levels of human Fis1 at the mitochondrial outer membrane regulate mitochondrial morphology. J Cell Sci 2004; 117(Pt 7):1201-1210.
92. Yoon Y, Krueger EW, Oswald BJ et al. The mitochondrial protein hFis1 regulates mitochondrial fission in mammalian cells through an interaction with the dynamin-like protein DLP1. Mol Cell Biol 2003; 23(15):5409-5420.
93. Cipolat S, Rudka T, Hartmann D et al. Mitochondrial rhomboid PARL regulates cytochrome c release during apoptosis via OPA1-dependent cristae remodeling. Cell 2006; 126(1):163-175.
94. Frezza C, Cipolat S, Martins de Brito O et al. OPA1 controls apoptotic cristae remodeling independently from mitochondrial fusion. Cell 2006; 126(1):177-189.

95. Hermann GJ, Thatcher JW, Mills JP et al. Mitochondrial fusion in yeast requires the transmembrane GTPase Fzo1p. J Cell Biol 1998; 143(2):359-373.

96. Koshiba T, Detmer SA, Kaiser JT et al. Structural basis of mitochondrial tethering by mitofusin complexes. Science 2004; 305(5685):858-862.

97. Rapaport D, Brunner M, Neupert W et al. Fzo1p is a mitochondrial outer membrane protein essential for the biogenesis of functional mitochondria in Saccharomyces cerevisiae. J Biol Chem 1998; 273(32):20150-20155.

98. Wong ED, Wagner JA, Gorsich SW et al. The dynamin-related GTPase, Mgm1p, is an intermembrane space protein required for maintenance of fusion competent mitochondria. J Cell Biol 2000; 151(2):341-352.

99. Desagher S, Martinou JC. Mitochondria as the central control point of apoptosis. Trends Cell Biol 2000; 10(9):369-377.

100. Frank S, Gaume B, Bergmann-Leitner ES et al. The role of dynamin-related protein 1, a mediator of mitochondrial fission, in apoptosis. Dev Cell 2001; 1(4):515-525.

101. Karbowski M, Lee YJ, Gaume B et al. Spatial and temporal association of Bax with mitochondrial fission sites, Drp1 and Mfn2 during apoptosis. J Cell Biol 2002; 159(6):931-938.

102. Cassidy-Stone A, Chipuk JE, Ingerman E et al. Chemical inhibition of the mitochondrial division dynamin reveals its role in Bax/Bak-dependent mitochondrial outer membrane permeabilization. Dev Cell 2008; 14(2):193-204.

103. Goyal G, Fell B, Sarin A et al. Role of mitochondrial remodeling in programmed cell death in Drosophila melanogaster. Dev Cell 2007; 12(5):807-816.

104. Lee YJ, Jeong SY, Karbowski M et al. Roles of the mammalian mitochondrial fission and fusion mediators Fis1, Drp1 and Opa1 in apoptosis. Mol Biol Cell 2004; 15(11):5001-5011.

105. Parone PA, James DI, Da Cruz S et al. Inhibiting the mitochondrial fission machinery does not prevent Bax/Bak-dependent apoptosis. Mol Cell Biol 2006; 26(20):7397-7408.

106. Neuspiel M, Zunino R, Gangaraju S et al. Activated mitofusin 2 signals mitochondrial fusion, interferes with Bax activation and reduces susceptibility to radical induced depolarization. J Biol Chem 2005; 280(26):25060-25070.

107. Wasiak S, Zunino R, McBride HM. Bax/Bak promote sumoylation of DRP1 and its stable association with mitochondria during apoptotic cell death. J Cell Biol 2007; 177(3):439-450.

108. Niemann A, Ruegg M, La Padula V et al. Ganglioside-induced differentiation associated protein 1 is a regulator of the mitochondrial network: new implications for Charcot-Marie-Tooth disease. J Cell Biol 2005; 170(7):1067-1078.

109. Szabadkai G, Simoni AM, Chami M et al. Drp-1-dependent division of the mitochondrial network blocks intraorganellar Ca^{2+} waves and protects against Ca^{2+}-mediated apoptosis. Mol Cell 2004; 16(1):59-68.

110. Tondera D, Czauderna F, Paulick K et al. The mitochondrial protein MTP18 contributes to mitochondrial fission in mammalian cells. J Cell Sci 2005; 118(Pt 14):3049-3059.

111. Tondera D, Santel A, Schwarzer R et al. Knockdown of MTP18, a novel phosphatidylinositol 3-kinase-dependent protein, affects mitochondrial morphology and induces apoptosis. J Biol Chem 2004; 279(30):31544-31555.

112. Taguchi N, Ishihara N, Jofuku A et al. Mitotic phosphorylation of dynamin-related GTPase Drp1 participates in mitochondrial fission. J Biol Chem 2007; 282(15):11521-11529.

113. Szabadkai G, Simoni AM, Bianchi K et al. Mitochondrial dynamics and Ca^{2+} signaling. Biochim Biophys Acta 2006; 1763(5-6):442-449.

114. Brooks C, Wei Q, Feng L et al. Bak regulates mitochondrial morphology and pathology during apoptosis by interacting with mitofusins. Proc Natl Acad Sci USA 2007; 104(28):11649-11654.

115. Karbowski M, Norris KL, Cleland MM et al. Role of Bax and Bak in mitochondrial morphogenesis. Nature 2006; 443(7112):658-662.

116. Delivani P, Adrain C, Taylor RC et al. Role for CED-9 and Egl-1 as regulators of mitochondrial fission and fusion dynamics. Mol Cell 2006; 21(6):761-773.

117. Li H, Chen Y, Jones AF et al. Bcl-xL induces Drp1-dependent synapse formation in cultured hippocampal neurons. Proc Natl Acad Sci USA 2008; 105(6):2169-2174.

118. Li Z, Okamoto K, Hayashi Y et al. The importance of dendritic mitochondria in the morphogenesis and plasticity of spines and synapses. Cell 2004; 119(6):873-887.

119. Verstreken P, Ly CV, Venken KJ et al. Synaptic mitochondria are critical for mobilization of reserve pool vesicles at Drosophila neuromuscular junctions. Neuron 2005; 47(3):365-378.

120. Berman SB, Chen YB, Qi B et al. Bcl-x L increases mitochondrial fission, fusion and biomass in neurons. J Cell Biol 2009; 184(5):707-719.

121. Breckenridge DG, Kang BH, Xue D. BCL-2 Proteins EGL-1 and CED-9 Do Not Regulate Mitochondrial Fission or Fusion in Caenorhabditis elegans. Curr Biol 2009.

122. Jagasia R, Grote P, Westermann B et al. DRP-1-mediated mitochondrial fragmentation during EGL-1-induced cell death in C. elegans. Nature 2005; 433(7027):754-760.
123. Danial NN, Gramm CF, Scorrano L et al. BAD and glucokinase reside in a mitochondrial complex that integrates glycolysis and apoptosis. Nature 2003; 424(6951):952-956.
124. Matschinsky FM, Magnuson MA, Zelent D et al. The network of glucokinase-expressing cells in glucose homeostasis and the potential of glucokinase activators for diabetes therapy. Diabetes 2006; 55(1):1-12.
125. Danial NN, Walensky LD, Zhang CY et al. Dual role of proapoptotic BAD in insulin secretion and beta cell survival. Nat Med 2008; 14(2):144-153.
126. Arden C, Baltrusch S, Agius L. Glucokinase regulatory protein is associated with mitochondria in hepatocytes. FEBS Lett 2006; 580(8):2065-2070.
127. Wilson JE. Isozymes of mammalian hexokinase: structure, subcellular localization and metabolic function. J Exp Biol 2003; 206(Pt 12):2049-2057.
128. Wiederkehr A, Wollheim CB. Minireview: implication of mitochondria in insulin secretion and action. Endocrinology 2006; 147(6):2643-2649.
129. Balasubramanian R, Karve A, Kandasamy M et al. A role for F-actin in hexokinase-mediated glucose signaling. Plant Physiol 2007; 145(4):1423-1434.
130. Giege P, Heazlewood JL, Roessner-Tunali U et al. Enzymes of glycolysis are functionally associated with the mitochondrion in Arabidopsis cells. Plant Cell 2003; 15(9):2140-2151.
131. Murata T, Katagiri H, Ishihara H et al. Colocalization of glucokinase with actin filaments. FEBS Lett 1997; 406(1-2):109-113.
132. Waingeh VF, Gustafson CD, Kozliak EI et al. Glycolytic enzyme interactions with yeast and skeletal muscle F-actin. Biophys J 2006; 90(4):1371-1384.
133. Zhou YP, Pena JC, Roe MW et al. Overexpression of Bcl-x(L) in beta-cells prevents cell death but impairs mitochondrial signal for insulin secretion. Am J Physiol Endocrinol Metab 2000; 278(2):E340-351.
134. Walensky LD, Pitter K, Morash J et al. A stapled BID BH3 helix directly binds and activates BAX. Mol Cell 2006; 24(2):199-210.
135. Bose AK, Mocanu MM, Carr RD et al. Glucagon-like peptide 1 can directly protect the heart against ischemia/reperfusion injury. Diabetes 2005; 54(1):146-151.
136. Liu W, Chin-Chance C, Lee EJ et al. Activation of phosphatidylinositol 3-kinase contributes to insulin-like growth factor I-mediated inhibition of pancreatic beta-cell death. Endocrinology 2002; 143(10):3802-3812.
137. Danial NN. BAD: undertaker by night, candyman by day Oncogene. 2009; in press.
138. Harada H, Andersen JS, Mann M et al. p70S6 kinase signals cell survival as well as growth, inactivating the pro-apoptotic molecule BAD. Proc Natl Acad Sci USA 2001; 98(17):9666-9670.
139. Tan Y, Demeter MR, Ruan H et al. BAD Ser-155 phosphorylation regulates BAD/BCL-XL interaction and cell survival. J Biol Chem 2000; 275(33):25865-25869.
140. Yu C, Minemoto Y, Zhang J et al. JNK suppresses apoptosis via phosphorylation of the proapoptotic BCL-2 family protein BAD. Mol Cell 2004; 13(3):329-340.
141. Deng H, Yu F, Chen J et al. Phosphorylation of Bad at Thr-201 by JNK1 promotes glycolysis through activation of phosphofructokinase-1. J Biol Chem 2008; 283(30):20754-20760.
142. Smith WE, Langer S, Wu C et al. Molecular coordination of hepatic glucose metabolism by the 6-phosphofructo-2-kinase/fructose-2,6-bisphosphatase:glucokinase complex. Mol Endocrinol 2007; 21(6):1478-1487.
143. Karantza-Wadsworth V, Patel S, Kravchuk O et al. Autophagy mitigates metabolic stress and genome damage in mammary tumorigenesis. Genes Dev 2007; 21(13):1621-1635.
144. Levine B, Kroemer G. Autophagy in the pathogenesis of disease. Cell 2008; 132(1):27-42.
145. Fimia GM, Stoykova A, Romagnoli A et al. Ambra1 regulates autophagy and development of the nervous system. Nature 2007; 447(7148):1121-1125.
146. Liang C, Feng P, Ku B et al. Autophagic and tumour suppressor activity of a novel Beclin1-binding protein UVRAG. Nat Cell Biol 2006; 8(7):688-699.
147. Takahashi Y, Coppola D, Matsushita N et al. Bif-1 interacts with Beclin 1 through UVRAG and regulates autophagy and tumorigenesis. Nat Cell Biol 2007; 9(10):1142-1151.
148. Levine B, Klionsky DJ. Development by self-digestion: molecular mechanisms and biological functions of autophagy. Dev Cell 2004; 6(4):463-477.
149. Geng J, Klionsky DJ. The Atg8 and Atg12 ubiquitin-like conjugation systems in macroautophagy. 'Protein modifications: beyond the usual suspects' review series. EMBO Rep 2008; 9(9):859-864.
150. Lum JJ, DeBerardinis RJ, Thompson CB. Autophagy in metazoans: cell survival in the land of plenty. Nat Rev Mol Cell Biol 2005; 6(6):439-448.
151. Sarbassov DD, Ali SM, Sabatini DM. Growing roles for the mTOR pathway. Curr Opin Cell Biol 2005; 17(6):596-603.

152. Diaz-Troya S, Perez-Perez ME, Florencio FJ et al. The role of TOR in autophagy regulation from yeast to plants and mammals. Autophagy 2008; 4(7):851-865.
153. Lum JJ, Bauer DE, Kong M et al. Growth factor regulation of autophagy and cell survival in the absence of apoptosis. Cell 2005; 120(2):237-248.
154. Levine B, Yuan J. Autophagy in cell death: an innocent convict? J Clin Invest 2005; 115(10):2679-2688.
155. Lockshin RA, Zakeri Z. Apoptosis, autophagy and more. Int J Biochem Cell Biol 2004; 36(12):2405-2419.
156. Kroemer G, Levine B. Autophagic cell death: the story of a misnomer. Nat Rev Mol Cell Biol 2008; 9(12):1004-1010.
157. Liang XH, Kleeman LK, Jiang HH et al. Protection against fatal Sindbis virus encephalitis by beclin, a novel BCL-2-interacting protein. J Virol 1998; 72(11):8586-8596.
158. Pattingre S, Tassa A, Qu X et al. BCL-2 antiapoptotic proteins inhibit Beclin 1-dependent autophagy. Cell 2005; 122(6):927-939.
159. Maiuri MC, Le Toumelin G, Criollo A et al. Functional and physical interaction between Bcl-X(L) and a BH3-like domain in Beclin-1. EMBO J 2007; 26(10):2527-2539.
160. Oberstein A, Jeffrey PD, Shi Y. Crystal structure of the BCL-XL-Beclin 1 peptide complex: Beclin 1 is a novel BH3-only protein. J Biol Chem 2007; 282(17):13123-13132.
161. Erlich S, Mizrachy L, Segev O et al. Differential interactions between Beclin 1 and BCL-2 family members. Autophagy 2007; 3(6):561-568.
162. Liang XH, Jackson S, Seaman M et al. Induction of autophagy and inhibition of tumorigenesis by beclin 1. Nature 1999; 402(6762):672-676.
163. Wei Y, Pattingre S, Sinha S et al. JNK1-mediated phosphorylation of BCL-2 regulates starvation-induced autophagy. Mol Cell 2008; 30(6):678-688.
164. Abedin MJ, Wang D, McDonnell MA et al. Autophagy delays apoptotic death in breast cancer cells following DNA damage. Cell Death Differ 2007; 14(3):500-510.
165. Daido S, Kanzawa T, Yamamoto A et al. Pivotal role of the cell death factor BNIP3 in ceramide-induced autophagic cell death in malignant glioma cells. Cancer Res 2004; 64(12):4286-4293.
166. Hamacher-Brady A, Brady NR, Logue SE et al. Response to myocardial ischemia/reperfusion injury involves Bnip3 and autophagy. Cell Death Differ 2007; 14(1):146-157.
167. Rashmi R, Pillai SG, Vijayalingam S et al. BH3-only protein BIK induces caspase-independent cell death with autophagic features in BCL-2 null cells. Oncogene 2008; 27(10):1366-1375.
168. Tracy K, Dibling BC, Spike BT et al. BNIP3 is an RB/E2F target gene required for hypoxia-induced autophagy. Mol Cell Biol 2007; 27(17):6229-6242.
169. Zhang J, Ney PA. NIX induces mitochondrial autophagy in reticulocytes. Autophagy 2008; 4(3):354-356.
170. Oltersdorf T, Elmore SW, Shoemaker AR et al. An inhibitor of BCL-2 family proteins induces regression of solid tumours. Nature 2005; 435(7042):677-681.
171. Germain M, Mathai JP, Shore GC. BH-3-only BIK functions at the endoplasmic reticulum to stimulate cytochrome c release from mitochondria. J Biol Chem 2002; 277(20):18053-18060.
172. Criollo A, Maiuri MC, Tasdemir E et al. Regulation of autophagy by the inositol trisphosphate receptor. Cell Death Differ 2007; 14(5):1029-1039.
173. Hoyer-Hansen M, Bastholm L, Szyniarowski P et al. Control of macroautophagy by calcium, calmodulin-dependent kinase kinase-beta and BCL-2. Mol Cell 2007; 25(2):193-205.
174. Schweers RL, Zhang J, Randall MS et al. NIX is required for programmed mitochondrial clearance during reticulocyte maturation. Proc Natl Acad Sci USA 2007; 104(49):19500-19505.
175. Sandoval H, Thiagarajan P, Dasgupta SK et al. Essential role for Nix in autophagic maturation of erythroid cells. Nature 2008; 454(7201):232-235.
176. Heynen MJ, Tricot G, Verwilghen RL. Autophagy of mitochondria in rat bone marrow erythroid cells. Relation to nuclear extrusion. Cell Tissue Res 1985; 239(1):235-239.
177. Koury MJ, Koury ST, Kopsombut P et al. In vitro maturation of nascent reticulocytes to erythrocytes. Blood 2005; 105(5):2168-2174.
178. Waugh RE, McKenney JB, Bauserman RG et al. Surface area and volume changes during maturation of reticulocytes in the circulation of the baboon. J Lab Clin Med 1997; 129(5):527-535.
179. Ron D, Walter P. Signal integration in the endoplasmic reticulum unfolded protein response. Nat Rev Mol Cell Biol 2007; 8(7):519-529.
180. Scheuner D, Kaufman RJ. The unfolded protein response: a pathway that links insulin demand with beta-cell failure and diabetes. Endocr Rev 2008; 29(3):317-333.
181. Todd DJ, Lee AH, Glimcher LH. The endoplasmic reticulum stress response in immunity and autoimmunity. Nat Rev Immunol 2008; 8(9):663-674.

182. Fujita E, Kouroku Y, Isoai A et al. Two endoplasmic reticulum-associated degradation (ERAD) systems for the novel variant of the mutant dysferlin: ubiquitin/proteasome ERAD(I) and autophagy/lysosome ERAD(II). Hum Mol Genet 2007; 16(6):618-629.

183. Ron D, Hampton RY. Membrane biogenesis and the unfolded protein response. J Cell Biol 2004; 167(1):23-25.

184. Berridge MJ. The endoplasmic reticulum: a multifunctional signaling organelle. Cell Calcium 2002; 32(5-6):235-249.

185. Sakaki K, Wu J, Kaufman RJ. Protein kinase Ctheta is required for autophagy in response to stress in the endoplasmic reticulum. J Biol Chem 2008; 283(22):15370-15380.

186. Szegezdi E, Logue SE, Gorman AM et al. Mediators of endoplasmic reticulum stress-induced apoptosis. EMBO Rep 2006; 7(9):880-885.

187. Futami T, Miyagishi M, Taira K. Identification of a network involved in thapsigargin-induced apoptosis using a library of small interfering RNA expression vectors. J Biol Chem 2005; 280(1):826-831.

188. Hetz C, Thielen P, Fisher J et al. The proapoptotic BCL-2 family member BIM mediates motoneuron loss in a model of amyotrophic lateral sclerosis. Cell Death Differ 2007; 14(7):1386-1389.

189. Li J, Lee B, Lee AS. Endoplasmic reticulum stress-induced apoptosis: multiple pathways and activation of p53-up-regulated modulator of apoptosis (PUMA) and NOXA by p53. J Biol Chem 2006; 281(11):7260-7270.

190. Luo X, He Q, Huang Y et al. Transcriptional upregulation of PUMA modulates endoplasmic reticulum calcium pool depletion-induced apoptosis via Bax activation. Cell Death Differ 2005; 12(10):1310-1318.

191. McCullough KD, Martindale JL, Klotz LO et al. Gadd153 sensitizes cells to endoplasmic reticulum stress by down-regulating Bcl2 and perturbing the cellular redox state. Mol Cell Biol 2001; 21(4):1249-1259.

192. Puthalakath H, O'Reilly LA, Gunn P et al. ER stress triggers apoptosis by activating BH3-only protein Bim. Cell 2007; 129(7):1337-1349.

193. Yamaguchi H, Wang HG. CHOP is involved in endoplasmic reticulum stress-induced apoptosis by enhancing DR5 expression in human carcinoma cells. J Biol Chem 2004; 279(44):45495-45502.

194. Ohoka N, Yoshii S, Hattori T et al. TRB3, a novel ER stress-inducible gene, is induced via ATF4-CHOP pathway and is involved in cell death. EMBO J 2005; 24(6):1243-1255.

195. Elyaman W, Terro F, Suen KC et al. BAD and BCL-2 regulation are early events linking neuronal endoplasmic reticulum stress to mitochondria-mediated apoptosis. Brain Res Mol Brain Res 2002; 109(1-2):233-238.

196. Upton JP, Austgen K, Nishino M et al. Caspase-2 cleavage of BID is a critical apoptotic signal downstream of endoplasmic reticulum stress. Mol Cell Biol 2008; 28(12):3943-3951.

197. Urano F, Wang X, Bertolotti A et al. Coupling of stress in the ER to activation of JNK protein kinases by transmembrane protein kinase IRE1. Science 2000; 287(5453):664-666.

198. Huang CJ, Haataja L, Gurlo T et al. Induction of endoplasmic reticulum stress-induced beta-cell apoptosis and accumulation of polyubiquitinated proteins by human islet amyloid polypeptide. Am J Physiol Endocrinol Metab 2007; 293(6):E1656-1662.

199. Nishitoh H, Matsuzawa A, Tobiume K et al. ASK1 is essential for endoplasmic reticulum stress-induced neuronal cell death triggered by expanded polyglutamine repeats. Genes Dev 2002; 16(11):1345-1355.

200. Lin JH, Li H, Yasumura D et al. IRE1 signaling affects cell fate during the unfolded protein response. Science 2007; 318(5852):944-949.

201. Lin JH, Li H, Zhang Y et al. Divergent effects of PERK and IRE1 signaling on cell viability. PLoS ONE 2009; 4(1):e4170.

202. Zinszner H, Kuroda M, Wang X et al. CHOP is implicated in programmed cell death in response to impaired function of the endoplasmic reticulum. Genes Dev 1998; 12(7):982-995.

203. Rutkowski DT, Arnold SM, Miller CN et al. Adaptation to ER stress is mediated by differential stabilities of pro-survival and pro-apoptotic mRNAs and proteins. PLoS Biol 2006; 4(11):e374.

204. Hetz C, Bernasconi P, Fisher J et al. Proapoptotic BAX and BAK modulate the unfolded protein response by a direct interaction with IRE1alpha. Science 2006; 312(5773):572-576.

205. Papa FR, Zhang C, Shokat K et al. Bypassing a kinase activity with an ATP-competitive drug. Science 2003; 302(5650):1533-1537.

206. Han D, Upton JP, Hagen A et al. A kinase inhibitor activates the IRE1alpha RNase to confer cytoprotection against ER stress. Biochem Biophys Res Commun 2008; 365(4):777-783.

207. Klee M, Pimentel-Muinos FX. Bcl-X(L) specifically activates Bak to induce swelling and restructuring of the endoplasmic reticulum. J Cell Biol 2005; 168(5):723-734.

208. Aylon Y, Oren M. Living with p53, dying of p53. Cell 2007; 130(4):597-600.

209. Hoffman B, Liebermann DA. Apoptotic signaling by c-MYC. Oncogene 2008; 27(50):6462-6472.

210. Iaquinta PJ, Lees JA. Life and death decisions by the E2F transcription factors. Curr Opin Cell Biol 2007; 19(6):649-657.
211. Altieri DC. The case for survivin as a regulator of microtubule dynamics and cell-death decisions. Curr Opin Cell Biol 2006; 18(6):609-615.
212. Lamkanfi M, Festjens N, Declercq W et al. Caspases in cell survival, proliferation and differentiation. Cell Death Differ 2007; 14(1):44-55.
213. Zinkel S, Gross A, Yang E. BCL2 family in DNA damage and cell cycle control. Cell Death Differ 2006; 13(8):1351-1359.
214. Borner C. Diminished cell proliferation associated with the death-protective activity of BCL-2. J Biol Chem 1996; 271(22):12695-12698.
215. Fujise K, Zhang D, Liu J et al. Regulation of apoptosis and cell cycle progression by MCL1. Differential role of proliferating cell nuclear antigen. J Biol Chem 2000; 275(50):39458-39465.
216. Jamil S, Sobouti R, Hojabrpour P et al. A proteolytic fragment of Mcl-1 exhibits nuclear localization and regulates cell growth by interaction with Cdk1. Biochem J 2005; 387(Pt 3):659-667.
217. Linette GP, Li Y, Roth K et al. Cross talk between cell death and cell cycle progression: BCL-2 regulates NFAT-mediated activation. Proc Natl Acad Sci USA 1996; 93(18):9545-9552.
218. Mazel S, Burtrum D, Petrie HT. Regulation of cell division cycle progression by bcl-2 expression: a potential mechanism for inhibition of programmed cell death. J Exp Med 1996; 183(5):2219-2226.
219. O'Reilly LA, Huang DC, Strasser A. The cell death inhibitor BCL-2 and its homologues influence control of cell cycle entry. EMBO J 1996; 15(24):6979-6990.
220. Greider C, Chattopadhyay A, Parkhurst C et al. BCL-x(L) and BCL2 delay Myc-induced cell cycle entry through elevation of p27 and inhibition of G1 cyclin-dependent kinases. Oncogene 2002; 21(51):7765-7775.
221. Lind EF, Wayne J, Wang QZ et al. BCL-2-induced changes in E2F regulatory complexes reveal the potential for integrated cell cycle and cell death functions. J Immunol 1999; 162(9):5374-5379.
222. Trimarchi JM, Lees JA. Sibling rivalry in the E2F family. Nat Rev Mol Cell Biol 2002; 3(1):11-20.
223. Vairo G, Soos TJ, Upton TM et al. BCL-2 retards cell cycle entry through p27(Kip1), pRB relative p130 and altered E2F regulation. Mol Cell Biol 2000; 20(13):4745-4753.
224. Jones RG, Bui T, White C et al. The proapoptotic factors Bax and Bak regulate T-Cell proliferation through control of endoplasmic reticulum Ca(2+) homeostasis. Immunity 2007; 27(2):268-280.
225. Janumyan Y, Cui Q, Yan L et al. G0 function of BCL2 and BCL-xL requires BAX, BAK and p27 phosphorylation by Mirk, revealing a novel role of BAX and BAK in quiescence regulation. J Biol Chem 2008; 283(49):34108-34120.
226. Brady HJ, Gil-Gomez G, Kirberg J et al. Bax alpha perturbs T-cell development and affects cell cycle entry of T-cells. EMBO J 1996; 15(24):6991-7001.
227. Deng X, Mercer SE, Shah S et al. The cyclin-dependent kinase inhibitor p27Kip1 is stabilized in G(0) by Mirk/dyrk1B kinase. J Biol Chem 2004; 279(21):22498-22504.
228. Chattopadhyay A, Chiang CW, Yang E. BAD/BCL-[X(L)] heterodimerization leads to bypass of G0/G1 arrest. Oncogene 2001; 20(33):4507-4518.
229. Janumyan YM, Sansam CG, Chattopadhyay A et al. Bcl-xL/BCL-2 coordinately regulates apoptosis, cell cycle arrest and cell cycle entry. EMBO J 2003; 22(20):5459-5470.
230. Mok CL, Gil-Gomez G, Williams O et al. Bad can act as a key regulator of T-cell apoptosis and T-cell development. J Exp Med 1999; 189(3):575-586.
231. Uhlmann EJ, D'Sa-Eipper C, Subramanian T et al. Deletion of a nonconserved region of BCL-2 confers a novel gain of function: suppression of apoptosis with concomitant cell proliferation. Cancer Res 1996; 56(11):2506-2509.
232. Huang LJ, Durick K, Weiner JA et al. Identification of a novel protein kinase A anchoring protein that binds both type I and type II regulatory subunits. J Biol Chem 1997; 272(12):8057-8064.
233. Cheng N, Janumyan YM, Didion L et al. BCL-2 inhibition of T-cell proliferation is related to prolonged T-cell survival. Oncogene 2004; 23(21):3770-3780.
234. Pinton P, Rizzuto R. BCL-2 and Ca²⁺ homeostasis in the endoplasmic reticulum. Cell Death Differ 2006; 13(8):1409-1418.
235. Gottlieb E, Vander Heiden MG, Thompson CB. Bcl-x(L) prevents the initial decrease in mitochondrial membrane potential and subsequent reactive oxygen species production during tumor necrosis factor alpha-induced apoptosis. Mol Cell Biol 2000; 20(15):5680-5689.
236. Hockenbery DM, Oltvai ZN, Yin XM et al. BCL-2 functions in an antioxidant pathway to prevent apoptosis. Cell 1993; 75(2):241-251.
237. Mandal S, Guptan P, Owusu-Ansah E et al. Mitochondrial regulation of cell cycle progression during development as revealed by the tenured mutation in Drosophila. Dev Cell 2005; 9(6):843-854.
238. Owusu-Ansah E, Yavari A, Mandal S et al. Distinct mitochondrial retrograde signals control the G1-S cell cycle checkpoint. Nat Genet 2008; 40(3):356-361.

239. Hardie DG. AMP-activated/SNF1 protein kinases: conserved guardians of cellular energy. Nat Rev Mol Cell Biol 2007; 8(10):774-785.
240. Imamura K, Ogura T, Kishimoto A et al. Cell cycle regulation via p53 phosphorylation by a 5'-AMP activated protein kinase activator, 5-aminoimidazole- 4-carboxamide-1-beta-D-ribofuranoside, in a human hepatocellular carcinoma cell line. Biochem Biophys Res Commun 2001; 287(2):562-567.
241. Jones RG, Plas DR, Kubek S et al. AMP-activated protein kinase induces a p53-dependent metabolic checkpoint. Mol Cell 2005; 18(3):283-293.
242. Liang J, Shao SH, Xu ZX et al. The energy sensing LKB1-AMPK pathway regulates p27(kip1) phosphorylation mediating the decision to enter autophagy or apoptosis. Nat Cell Biol 2007; 9(2):218-224.
243. Devadas S, Zaritskaya L, Rhee SG et al. Discrete generation of superoxide and hydrogen peroxide by T-cell receptor stimulation: selective regulation of mitogen-activated protein kinase activation and fas ligand expression. J Exp Med 2002; 195(1):59-70.
244. Haughn L, Hawley RG, Morrison DK et al. BCL-2 and BCL-XL restrict lineage choice during hematopoietic differentiation. J Biol Chem 2003; 278(27):25158-25165.
245. Shibasaki F, Kondo E, Akagi T et al. Suppression of signalling through transcription factor NF-AT by interactions between calcineurin and BCL-2. Nature 1997; 386(6626):728-731.
246. Portier BP, Taglialatela G. BCL-2 localized at the nuclear compartment induces apoptosis after transient overexpression. J Biol Chem 2006; 281(52):40493-40502.
247. Fernandez-Sarabia MJ, Bischoff JR. BCL-2 associates with the ras-related protein R-ras p23. Nature 1993; 366(6452):274-275.
248. Chen DF, Schneider GE, Martinou JC et al. BCL-2 promotes regeneration of severed axons in mammalian CNS. Nature 1997; 385(6615):434-439.
249. Hilton M, Middleton G, Davies AM. BCL-2 influences axonal growth rate in embryonic sensory neurons. Curr Biol 1997; 7(10):798-800.
250. Sato T, Hanada M, Bodrug S et al. Interactions among members of the BCL-2 protein family analyzed with a yeast two-hybrid system. Proc Natl Acad Sci USA 1994; 91(20):9238-9242.
251. Zhang KZ, Westberg JA, Holtta E et al. BCL2 regulates neural differentiation. Proc Natl Acad Sci USA 1996; 93(9):4504-4508.
252. Jiao J, Huang X, Feit-Leithman RA et al. BCL-2 enhances Ca(2+) signaling to support the intrinsic regenerative capacity of CNS axons. EMBO J 2005; 24(5):1068-1078.
253. Gotz R, Wiese S, Takayama S et al. Bag1 is essential for differentiation and survival of hematopoietic and neuronal cells. Nat Neurosci 2005; 8(9):1169-1178.
254. Planchamp V, Bermel C, Tonges L et al. BAG1 promotes axonal outgrowth and regeneration in vivo via Raf-1 and reduction of ROCK activity. Brain 2008; 131(Pt 10):2606-2619.
255. Moldovan GL, Pfander B, Jentsch S. PCNA, the maestro of the replication fork. Cell 2007; 129(4):665-679.
256. Bartek J, Lukas C, Lukas J. Checking on DNA damage in S phase. Nat Rev Mol Cell Biol 2004; 5(10):792-804.
257. Shiloh Y. The ATM-mediated DNA-damage response: taking shape. Trends Biochem Sci 2006; 31(7):402-410.
258. Kamer I, Sarig R, Zaltsman Y et al. Proapoptotic BID is an ATM effector in the DNA-damage response. Cell 2005; 122(4):593-603.
259. Song G, Chen GG, Chau DK et al. Bid exhibits S phase checkpoint activation and plays a pro-apoptotic role in response to etoposide-induced DNA damage in hepatocellular carcinoma cells. Apoptosis 2008; 13(5):693-701.
260. Zinkel SS, Hurov KE, Ong C et al. A role for proapoptotic BID in the DNA-damage response. Cell 2005; 122(4):579-591.
261. Grallert B, Boye E. The multiple facets of the intra-S checkpoint. Cell Cycle 2008; 7(15):2315-2320.

CHAPTER 2

Alternative Functions of the BCL-2 Protein Family at the Endoplasmic Reticulum

Diego Rojas-Rivera, Benjamin Caballero, Sebastian Zamorano, Fernanda Lisbona and Claudio Hetz*

Abstract

Apoptosis is essential for maintenance of tissue homeostasis and its deregulation results in a variety of disease conditions. The BCL-2 family of proteins is a group of evolutionarily conserved regulators of cell death that comprises both anti- and pro-apoptotic members, that operate at the mitochondrial membrane to control caspase activation. Different BCL-2-related proteins are also located in the endoplasmic reticulum (ER), where important roles in organelle physiology are proposed. Adaptation to ER stress is mediated by the activation of a complex signal transduction pathway known as the unfolded protein response (UPR). Recent reports indicate that the ER stress sensor IRE1α, signals through the formation of a protein complex platform at the ER membrane, here termed the "*UPRosome*". Alternatively, BCL-2 family members are contained in other multiprotein complexes at the ER that are involved in the control of diverse cellular processes including calcium homeostasis, autophagy and ER morphogenesis. Here we describe the emerging concept that BCL-2 family members are important regulators of essential cellular processes beyond apoptosis.

Introduction

Complex signaling responses mediate adaptation to organelle stress or initiation of apoptosis when a critical threshold of damage has been reached.[1] Execution of apoptosis depends on the activation of caspases, a process tightly regulated by the BCL-2 family of proteins. The BCL-2 family of proteins is comprised of pro- and anti-apoptotic members that are defined by the presence of up to four conserved domains. Anti-apoptotic BCL-2 family members display sequence homology in four α-helical domains called BCL-2 homology BH1 to BH4.[2] Pro-apoptotic BCL-2 members can be further subdivided into more highly conserved, "multidomain" members displaying homology in the BH1, BH2 and BH3 (i.e., BAX and BAK) domains, or the "BH3-only" members (i.e., BID, BIK, BIM, PUMA and NOXA)[3] which contain a single α-helical domain critical for activation of apoptosis. This pathway has gained in complexity since the identification of a new subgroup of pro-apoptotic proteins, called BNip proteins, that have minimal sequence similarity in the BH3 domain (reviewed in ref. 4).

*Corresponding Author: Claudio Hetz—Program of Cellular and Molecular Biology, The FONDAP Center for Molecular Studies of the Cell, Institute of Biomedical Sciences, Faculty of Medicine, and Millennium Nucleus for Neural Morphogenesis (NEMO), University of Chile, Santiago, Chile, P.O. BOX 70086. Website: http://ecb_icbm.med.uchile.cl/ Email: chetz@med.uchile.cl, chetz@hsph.harvard.edu

BCL-2 Protein Family: Essential Regulators of Cell Death, edited by Claudio Hetz.
©2010 Landes Bioscience and Springer Science+Business Media.

Most members of the BCL-2 family has distinct patterns of developmental expression, subcellular localization and differential responsiveness to specific death stimuli (reviewed in ref. 5). Some BH3-only proteins are thought to operate as sentinels of cellular damage,[2] where in response to various death stimuli (i.e., oxidative stress, growth factor deprivation, DNA damage, death receptor engagement, etc), they are activated either by transcriptional upregulation or through posttranslational modifications. BH3-only proteins then promote the activation of the core pro-apoptotic components BAX and/or BAK.[6] Activation of BAX and BAK is mediated by their intramembranous oligomerization and resultant permeabilization of the mitochondrial outer membrane.[2] Released mitochondrial proteins, such as cytochrome *c*, initiate caspase-mediated cell death.[7]

Accumulating evidence indicates that members of all three subclasses of the BCL-2 family of proteins are also located at the endoplasmic reticulum (ER) membrane.[8] The ER is an organelle with multiple functions. It is the site where the biosynthesis of steroids, cholesterol and other lipids occurs and hence plays a crucial role in organelle biogenesis and signaling by the generation of lipid second messengers. The two major roles of the ER are calcium storage and protein folding. Membrane-spanning and secreted proteins are synthesized in the ER and undergo posttranslational modifications, folding and proper oligomerization. A number of conditions (such as proteasome inhibition, mutant protein expression, ER calcium depletion, redox changes, etc) interfere with oxidative protein folding at the ER lumen,[9] resulting in the accumulation of unfolded or misfolded intermediates, a cellular condition referred to as "ER stress".

To alleviate ER stress cells activate a complex signaling pathway known as "unfolded protein response" (UPR). The UPR transmits information about the protein folding status in the ER lumen to the cytoplasm and the nucleus to decrease the unfolded protein load. This affects the expression of proteins involved in nearly every aspect of the secretory pathway, including protein entry into the ER, folding, glycosylation, ER-associated degradation, ER biogenesis and vesicular trafficking (reviewed in refs. 9, 10). However, if all these mechanisms of survival are insufficient to decrease the unfolded protein load, cells enter into apoptosis. Therefore, improper handling of ER stress constitutes a threat to cell life. The UPR/ER stress pathway has been implicated in many diseases, including neurodegenerative conditions,[11] cancer[12] and diabetes.[13] Increasing evidence suggests the BCL-2 family of proteins play a key role at the ER membrane were they modulate essential signaling processes in addition to cell death.[14] In this chapter, we focus on the alternative functions of the BCL-2 family of proteins at the ER membrane and their role in mediating the cellular responses against various types of cellular stress.

The BCL-2 Protein Family and the Unfolded Protein Response

The UPR was first characterized in yeast where a single signaling pathway, mediated by a type I transmembrane ER protein known as IRE1α (inositol-requiring transmembrane kinase/endonuclease) governs the response to ER stress.[15-18] In higher eukaryotes, the UPR is a more complex pathway mediated by at least three distinct UPR signaling pathways initiated by the sensors IRE1α, PERK (PKR-like ER kinase) and ATF6 (activating transcription factor 6). IRE1α is a Ser/Thr protein kinase and endoribonuclease that, upon activation, initiates the unconventional splicing of the mRNA encoding the transcription factor X-Box-binding protein 1 (XBP-1).[19-21] In mammalian cells, a 26 nucleotides intron of *xbp-1* mRNA is spliced out by activated IRE1α, leading to a shift in the coding reading frame. This splicing event promotes the expression of a more stable and potent transcriptional activator called XBP-1s that controls the upregulation of a broad spectrum of UPR-related genes involved in protein folding, redox metabolism, ER-associated degradation and protein quality control.[22] Indeed, XBP-1 expression is essential for the proper function of specialized secretory organs such as liver, pancreas and salivary gland in addition to B lymphocytes.[23-26]

IRE1α activation also controls the activation of the JNK (Jun N-terminal Kinase),[27,28] ERK (Extracellular signal-regulated kinase),[29] p38,[30] NF-κB (Nuclear factor kappa-light-chain-enhancer of activated B-cells)[31] and the ER-resident caspase 12; pathways which are controlled by distinct proteins. Different reports have shown that IRE1α activity is regulated by pro- and anti-apoptotic proteins through a protein platform, referred as the *UPRosome*.[32] For example, the pro-apoptotic protein AIP1 (ASK1-interacting protein 1) binds to IRE1α in response to ER stress and this interaction facilitates IRE1α dimerization in a kinase-independent manner. Interestingly, AIP1 specifically affected IRE1α activation and not PERK-dependent signaling.[33] TRAF2 (TNF-receptor associated factor-2) interacts in a kinase-dependent way with the cytosolic domain of active IRE1α. In consequence IRE1α, AIP1 and TRAF2 form a protein complex in response to ER stress, where AIP1 is necessary for the activation of JNK (c-Jun N-terminal kinase) and XBP-1 pathways.[33]

TRAF2 has been suggested to be linked to pro-caspase-12 processing and activation of NF-κB, JNK and ASK1. Under ER stress condition, an IRE1-TRAF2-ASK1 complex is formed to activate JNK signaling by ER stress.[28] TRAF2 also regulates NF-κB activation, mediating the formation of a complex between active IRE1α and inhibitor NF-κB kinase (IKK).[31] Contrary to the pro-survival effect of the IRE1α/XBP-1 pathway,[34] in general it seems that signals controlled by IRE1α-TRAF2 are strongly associated with cell death signals. However, the function of these UPR signaling branches is not well understood. In this context, activation of JNK by IRE1α has been proposed to activate apoptosis under irreversible ER stress condition in an analogous fashion to what has been described for the TNF receptor[35] but the mechanism linking these two phenomena is not clear.[36] The levels of the IRE1α signaling are also controlled by the ER-located PTP-1B (Protein-tyrosine phosphatase 1B). The absence of PTP-1B causes impaired JNK and p38 activation, XBP-1 splicing and expression of XBP-1 target genes. Similarly to AIP1, PTP-1B deficiency did not affect the activation of the stress sensor PERK.[30]

We recently provided evidence for a possible function of the BCL-2 protein family in the UPR (Fig. 1). BAX and BAK were shown to modulate the amplitude of IRE1α signaling.[37] BAX and BAK DKO cells displayed a specific deficiency in the auto-phosphorylation and oligomerization of IRE1α. BAX/BAK deficient mice and cells showed a decreased expression of IRE1α-downstream signals including JNK phosphorylation and XBP-1s expression under experimental ER stress conditions.[37] At the biochemical level, BAX and BAK form a protein complex with the cytosolic domain of IRE1α, possibly stabilizing its active form. The BH1 and BH3 domains were shown to be necessary for this regulatory activity. Overall, these findings suggested a new role for pro-apoptotic family members to act as accessory factors for the instigation of certain UPR signaling events. In this line, during early steps of UPR responses BAX and BAK may have pro-survival effects in adaptation to ER stress fomenting IRE1α signaling. Moreover, a recent report indicated that the specific expression of the BH3-only proteins BIM and PUMA at the ER leads to the activation of IRE1α/JNK pathway on a BAK-dependent manner.[38] Interestingly, these findings were obtained in the absence of any ER stressor, suggesting that these BH3-only proteins are potent activators of the IRE1α UPR branch. In this particular cell system, engagement of IRE1α/JNK was proposed to have a pro-apoptotic effect through a signaling to the mitochondria.[33]

BI-1 (Bax inhibitor 1) is an ER-resident protein that contains six transmembrane domains and interacts with different members of the BCL-2 family of proteins such as BCL-X$_L$ and BCL-2,[39] although it is functionally related to this family, it does not have apparent homology with their members. BI-1 is an evolutionary conserved protein described in eukaryotes, plants, bacteria and virus[40,41] and is present in species that do not have homologues of the BCL-2 family of proteins. Studies in human cells demonstrated that BI-1 overexpression protected from cell death induced by nutrient deprivation, DNA damage and oxidative stress, as BCL-2 overexpression does. BI-1 partially affects ER stress-dependent cell death in vivo.[42] Reed and colleagues described the ER stress responses of BI-1 knockout mice. BI-1 deficient mice showed hyper activation of the IRE1α pathway in vivo in a model of hepatic and renal ischemia.[43] These results were also recently recapitulated in cellular models of ER stress[44] where BI-1 overexpression decreased UPR activation under ER stress conditions as evidenced by decreased XBP-1s expression and JNK phosphorylation. These data suggested that BI-1 might have an inhibitory activity on the UPR, which contrasts with the opposite effect of BAX/BAK and BH3-only proteins on this pathway.

XBP-1 mRNA splicing levels decline after prolonged ER stress, whereas PERK signalling is sustained over time.[34] This mechanism may sensitize cells to apoptosis after chronic or irreversible

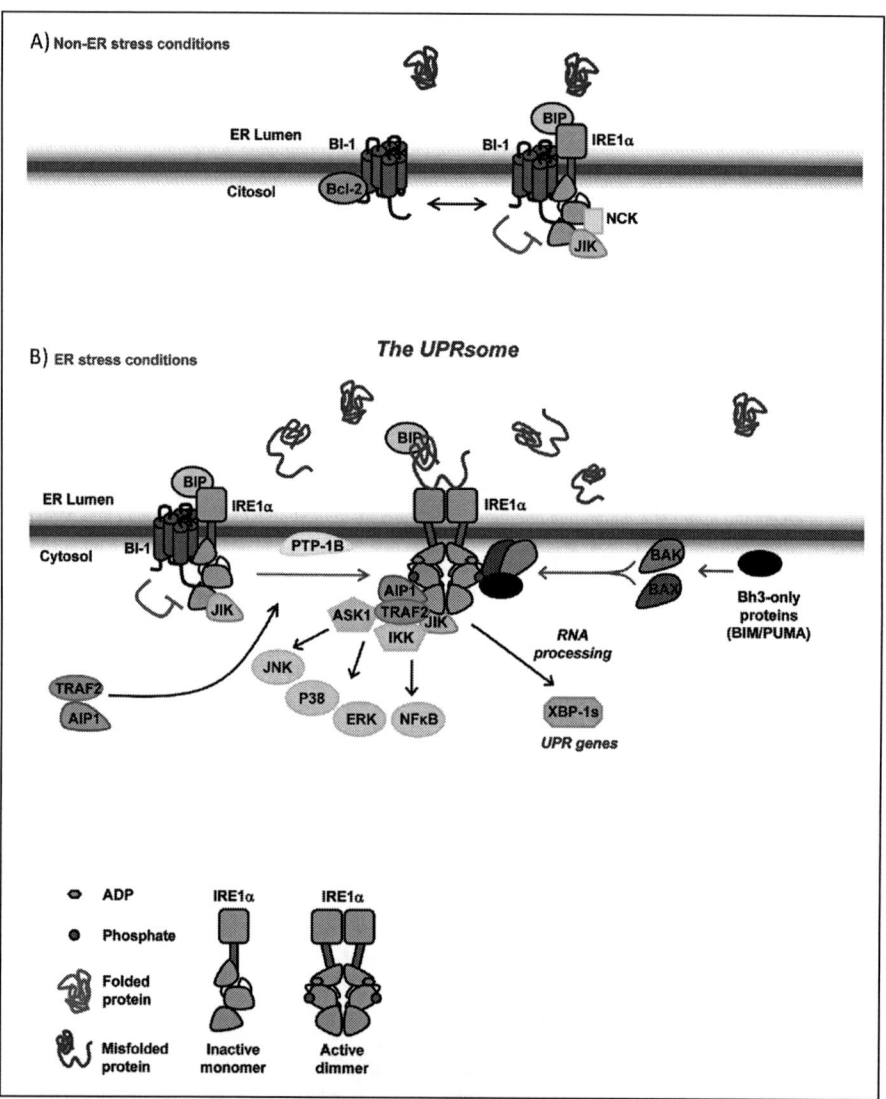

Figure 1. The *UPRosome*. Under nonstress conditions (A) IRE1α is maintained in an inactive monomeric state through its binding to the ER chaperone BiP in its lumenal domain and BI-1 in the cytosolic domain. Under ER stress condition (B), BiP binds preferentially to unfolded/ misfolded proteins at the ER lumen, releasing its partial inhibitory effect over IRE1α. The interaction between the ER lumenal domain of IRE1α with unfolded proteins may stabilize its dimeric state, leading to its autophosphorylation and to the activation of the RNAse. IRE1α signaling is mediated by the formation of a protein complex platform at the ER membrane, termed *UPRosome*. The *UPRosome* is constituted by multiple factors that modulate the activity of IRE1α and the initiation of a variety of signaling responses. Activation of IRE1α requires the binding of accessory proteins, such as PTP1B, AIP1, BAX and BAK. Active IRE1α interacts with AIP1 and TRAF2, initiating the activation of ASK1/JNK pathway. Sequestration of IKK by IRE1α/TRAF2 induces NF-κB signaling and TNF-alpha production. The ER located phosphatase PTP-1B is required for IRE1α activation. Finally, IRE1α may regulate the processing of pro-caspase-12 (in murine cells), through the binding of TRAF2.

ER stress, shutting down the pro-survival effects of IRE1α/XBP-1 signalling. However, it is not known how IRE1α is turned off. We recently reported that BI-1 has a crucial and direct role in modulating IRE1α function upstream of BAX and BAK. BI-1 expression suppressed IRE1α activity in fly and mouse models of ER stress. Increased levels of XBP-1s where observed in BI-1 KO mice under Tunicamycin (Tm) treatment compared to control animals. Flies overexpressing dBI-1 significantly decreased the levels of XBP-1 mRNA splicing in larvae exposed to Tm, indicating that BI-1 also regulates IRE1α in invertebrates. Of note, BI-1 deficient cells displayed hyperactivation of the ER stress sensor IRE1α, associated with a sustained splicing of XBP-1 over time and thus abrogating the inactivation phase of the stress sensor. These phenotypes were associated with the formation of a stable protein complex between BI-1 and IRE1α, decreasing its ribonuclease activity.[45] This model was reconstituted in vitro with purified components. BI-1 was initially cloned in yeast during a search for inhibitors that rescued the lethal phenotype of BAX overexpression.[39] Overexpression of BI-1 in yeast gives protection against heat shock and exposure to H_2O_2.[46] However, the biological function of BI-1 in yeast is not known, but it is an interesting candidate as a factor that might connect the UPR to the cell death process through evolution.

Our results, together with the aforementioned studies, suggest a model where the differential activation of ER stress sensors in different tissue contexts may be related to the engagement of specific regulatory complexes through the association of adaptor and direct binding of positive and negative modulators. We envision a model where a complex signalling platform is assembled at the level of IRE1α, the *UPRosome*, to modulate its activation status in terms of signalling intensity and kinetics of activation/inactivation.[32] In the context of cell survival, the fine tuning of UPR signalling is particularly relevant in life to death transitions by controlling transcriptional programs that regulate adaptation to stress or by initiating apoptosis of irreversibly damaged cells. The *UPRosome* concept may help to explain why each branch of the UPR signalize differentially in vivo, despite of operating through similar protein misfolding-sensing mechanism.[47] These findings suggest a model wherein the expression of anti- and pro-apoptotic proteins at the ER membrane may determine the amplitude of UPR responses and the ability of a cell to adapt to ER injuries. Furthermore, this model may give ideas to define a molecular switch to the homeostatic transition between adaptation and cell death.

Chronic ER Stress and BH3-Only Proteins

When the UPR cannot alleviate the overload of misfolded proteins in the ER, ER stress leads to cell death by apoptosis. Although the precise mechanism that triggers apoptosis by chronic ER stress is not clear, many proteins have been identified in the process,[36] among them, several BH3-only proteins (Fig. 2). This group of pro-apoptotic proteins has been mostly studied at the mitochondria, where they induce the activation of BAX and BAK leading to cytochrome c release and finally, apoptosis. The BH3-only proteins can be divided into two subtypes: activators (such as BID, BIM and possibly PUMA) and sensitizers (such as BAD and NOXA). Under physiological conditions, anti-apoptotic members of the BCL-2 family of proteins inhibit activator BH3-only proteins by a direct binding. When an apoptotic stimulus arrives, activators are further induced overcoming the inactivation by anti-apoptotic BCL-2 proteins. Alternatively, upregulation of sensitizers displace the activators from the anti-apoptotic proteins. Free activators then trigger BAX and BAK oligomerization and mitochondrial membrane permeabilization.[35,48-50]

BH3-only proteins can be transcriptional and posttranslational up-regulated. Two BH3-only proteins that undergo transcriptional up regulation under ER stress are PUMA and NOXA (Fig. 2). These proteins are mostly known as genes up-regulated by DNA damage, but under prolonged ER stress they are transcriptionaly induced in a p53-dependent manner.[51] In a pioneering study, cDNA microarray analysis showed that PUMA is up-regulated by ER stress.[52] Moreover, a global RNA interference screen for genes that regulate ER stress-mediated apoptosis corroborated the functional role of PUMA in this process.[53] In this study, NOXA was also identified as an apoptosis mediator. Additionally, MEFs cells lacking *puma* or *noxa* genes are partially resistant to ER stress-mediated apoptosis.[51] PUMA and NOXA are also up-regulated under ER stress conditions in a melanoma cells context.[54] Interestingly, in this study the anti-apoptotic protein MCL-1 (myeloid cell leukemia

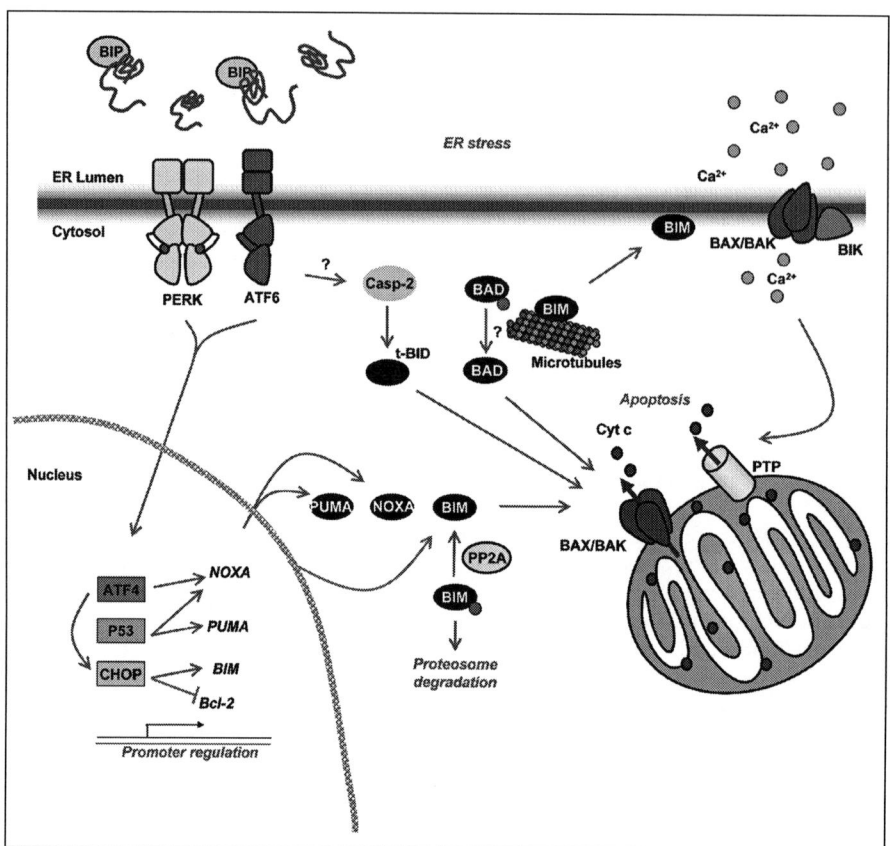

Figure 2. ER stress-mediated apoptosis. The BH3-only proteins have an important role in the regulation of ER stress-dependent apoptosis. PUMA and NOXA are transcriptionaly induced in cells undergoing ER stress in a p53-dependent manner. Additionally, BIM is induced and BCL-2 is down regulated by the UPR transcription factor CHOP, which is induced by the stress sensors PERK and ATF6. The unbalance between pro- and anti-apoptotic proteins leads to homo-oligomerization of BAX and/or BAK at the mitochondria with the consequent release of cytochrome c and caspases-dependent apoptosis. BIM is also posttranslationaly regulated by ER stress. BAD may also trigger apoptosis by its dephosphorylation under ER stress. BID activates apoptosis when is cleaved by caspase-2 under ER stress. When BIK, another BH3-only protein is over-expressed triggers BAX and BAK-dependent ER calcium release. Mitochondrial calcium uptake may trigger mitochondrial permeability transition pore (PTP), inducing the release of cytochrome c on a BAX/BAK-independent manner (see Fig. 3).

sequence 1) is also up regulated by ER stress and confers resistance to apoptosis (Fig. 2). Anticancer agents that induce ER stress such as Eeyarestatin I, lead to NOXA up-regulation by the ATF3 and ATF4 transcriptional factors.[55] Another BH3-only member, BIK, is primarily localized to the ER but it is neither transcriptionaly nor posttralationaly induced by ER stress.[56] BIK controls the release of calcium by BAK oligomerization at the ER triggering apoptosis.[57]

BIM is transcriptionaly and posttranslationaly up-regulated by ER stress (Fig. 2). BIM mRNA levels are induced by the transcription factor CHOP, an UPR-induced pro-apoptotic transcription factor. In addition, BCL-2 is downregulated by CHOP.[58] Moreover, BIM deficient mice are resistant to ER stress-induced apoptosis in vivo, similar to the phenotype described for CHOP deficient mice. Two different posttranslational mechanisms up-regulate BIM. Under physiological conditions, BIM is found in the dynein motor complex of the microtubule cytoskeleton and when ER stress is triggered, its translocated to the ER where it may promote caspase activation through an unknown mechanism.[59] In addition, dephosphorylation of BIM by the serine/threonine phosphatase 2A (PP2A) increases its pro-apoptotic activity under ER stress condition in different cell types, which prevents its ubiquitination and proteasomal degradation.[58] Correlative studies in rat primary cultures showed that BAD is also activated by dephosphorylation and produce apoptosis in cortical neurons undergoing ER stress.[60] Finally, another BH3-only protein, BID, is posttranslationaly up-regulated by caspase-2-dependent proteolytic activation upon ER-stress, leading to activation BAX/BAK at the mitochondria.[61]

The BCL-2 Protein Family and the Calcium Rheostat

One of the main known functions described for the BCL-2 family members at the ER is the control of calcium homeostasis (Fig. 3). The balance between anti- and pro-apoptotic proteins at the ER determines the steady state ER-calcium content possibly by modulating calcium leak. This activity has a direct impact on the amount of calcium released after stimulation. For example, DKO cells for BAX and BAK show decreased ER calcium content,[62,63] similar to the phenotype of BCL-2 overexpressing cells[64] and overexpression of different BH3-only proteins triggers calcium release (reviewed in refs. 8, 65, 66). Thus, the balance between different BCL-2 family of proteins at the ER may constitute a

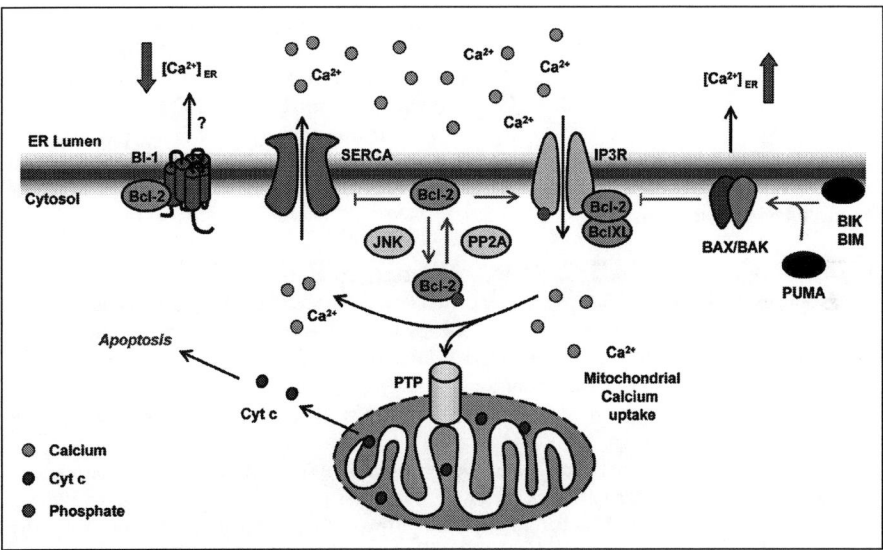

Figure 3. Regulation of ER calcium homeostasis by BCL-2 family members. Different anti- and pro-apoptotic members of the BCL-2 family of proteins are located at the ER membrane where they regulate ER calcium content. BCL-2 and BCL-X$_L$ interact with the IP3R calcium channel, modulating its activity. BCL-2 has been shown to increase ER calcium leak through the IP3R due to an increase of its phosphorylation. In contrast, BAX and BAK have the opposite effect on ER calcium content, a function that may be further modulated by BH3-only proteins (such as PUMA and BIK). BCL-2 reduces ER calcium content inactivating SERCA by a direct interaction. In addition, the activity of BCL-2 at the ER membrane is regulated by phosphorylation. JNK phosphorylates BCL-2, decreasing its anti-apoptotic activity and increasing ER calcium content, whereas the phosphatase PP2A decreases this phosphorylation through a direct interaction. BI-1 is also located at the ER membrane where it regulates calcium homeostasis reducing ER calcium content. On the other hand, calcium released from the ER influences mitochondrial-mediated apoptosis by membrane disruption through calcium uptake.

rheostat for the fine tuning of calcium metabolism. BI-1 expression also alters calcium homeostasis downstream of the BCL-2 family[67] where its expression reduces ER calcium content.[42,68] Interestingly, as in apoptosis BI-1 exerts an opposite effect to BAX/BAK in ER calcium content.

At the biochemical level, it has been documented by several groups that BCL-2 and BCL-X$_L$ form a protein complex with the IP3R (inositol triphosphate receptor),[69-71] modulating the channel opening (Fig. 3). The IP3R, together with the ryanodine receptor (RyR), are the main channels that controls ER calcium release in cells. Phosphorylation of BCL-2 by JNK in a non-structured loop has been shown to occur at the ER membrane. This modification also negatively regulates BCL-2's anti-apoptotic activity, in addition to its ability to bind to BH3-only proteins and control the ER calcium content.[64] The native protein complexes containing BCL-2 at the ER membrane were recently purified. BCL-2 was found to be directly regulated by a physical interaction with PP2A,[72] which dephosphorylates the sites targeted by JNK.

An inactivation and destabilization of the ER Ca^{2+} importer, sarcoplasmic/endoplasmic reticulum calcium ATPase (SERCA) is caused by its direct interaction with BCL-2,[73,74] resulting in reduced ER-calcium content. Finally, different BH3-only proteins are located at the ER or translocate to its membrane under stress conditions and impact calcium homeostasis (reviewed in ref. 8). Thus, one can speculate that depending on the cellular context and the stimuli, different BCL-2 containing protein complexes may exist at the ER membrane to control calcium signaling. Hence, the balance between pro- and anti-apoptotic BCL-2 related proteins at the ER determine the ER calcium content and rate of calcium release. The calcium uptake by the mitochondria has a direct impact on the susceptibility of a cell to apoptosis, influencing the drop in mitochondrial membrane potential and the release of cytochrome c and caspase activation (Fig. 3). Interestingly, on an elegant study, Scorrano and coworkers described that the control of ER calcium by BAX and BAK has a specific effect on calcium mediated-apoptosis (i.e., arachidonic acid, ceramide or H$_2$O$_2$ treatment) and not ER stress. On the other hand, reconstitution of BAX/BAK DKO cells with a mitochondrial-targeted BAX recovered the susceptibility of these cells to ER stress-induced apoptosis, without affecting the calcium phenotype. Thus, the control of calcium metabolism by the BCL-2 family probably reflects a new function beyond apoptosis.

Autophagy and the BCL-2 Protein Family of Proteins

Autophagy refers to the global process by which intracellular components are recycled through lysosome degradation (reviewed in ref. 75). Autophagy acts as a critical survival response under starvation conditions in which the degradation of intracellular proteins and organelles provides a source of amino acids during poor nutritional conditions. Intracellular components can be delivered to lysosomes for degradation by three different mechanisms known as macroautophagy, microautophagy and chaperone-mediated autophagy.[75] The best-studied form of autophagy is macroautophagy, hereafter referred to as autophagy. The hallmark of autophagy is the formation of double-membrane-bounded autophagosomes. Autophagosomes fuse with lysosomes to form autophagolysosomes, where intracellular components are degraded. Autophagy is a highly regulated process with complex steps that are controlled by a family of autophagic related genes (termed *atg* genes).[75,76] The generation of *atg* deficient mice unrevealed the function of autophagy in diverse processes, including development, cell differentiation, tissue remodeling, immunity, host-to-pathogen response and cell death/survival under stress conditions.[75] Beclin-1 (also known as Atg6) was the first identified mammalian autophagy gene product.[77] Beclin-1 is a haplo-insufficient tumor suppressor that was originally isolated as a BCL-2-interacting protein.[78-80] More interestingly, BCL-2 was recently shown to negatively regulate Beclin-1 through a direct binding.[81] Surprisingly, this regulatory activity of BCL-2 on autophagy was specifically attributed to its expression at the ER membrane,[81] suggesting that signaling events originating from the ER are crucial for autophagy (Fig.4). A recent report suggested a unidirectional regulation between BCL-2 and Beclin-1, since Beclin-1 binding to BCL-2 does not modify BCL-2 mediated apoptosis.[82] The formation of a BCL-2/Beclin-1 complex is also regulated by BH3-only proteins, revealing extensive crosstalk between apoptosis and autophagy.[83,84] More importantly, a

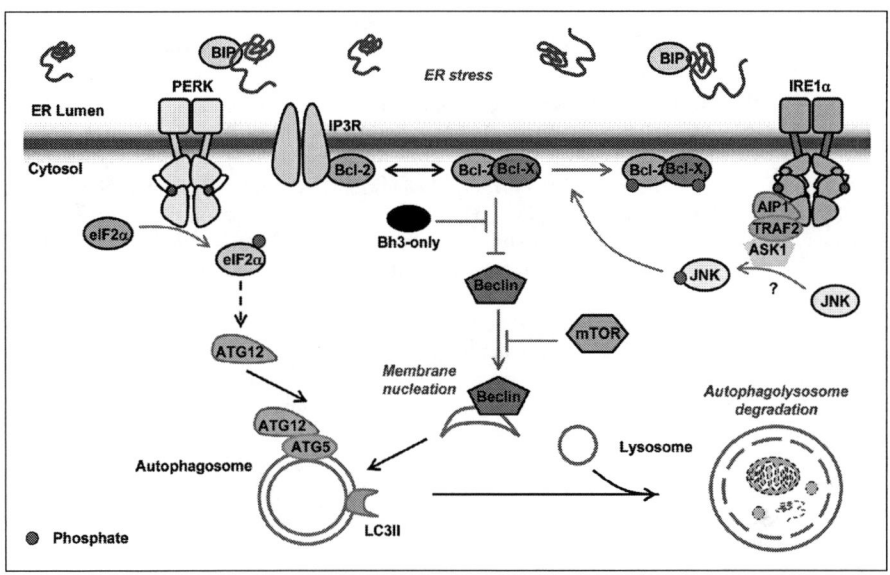

Figure 4. Regulation of autophagy by ER stress: Accumulation of misfolded proteins also triggers autophagy. Beclin-1 (Bec-1) is negatively regulated through an interaction with BCL-2 at the ER membrane, interaction that is antagonized by BH3-only proteins. On the other hand, IRE1α activation may increase the levels of autophagy through the activation of the JNK pathway. In addition, the IP3R controls autophagy that is dependent on BCL-2/BCL-X$_L$ expression. Activation of PERK may also potentate autophagy, through the transcriptional upregulation of Atg12. Bec-1 initiates membrane nucleation and autophagosome formation, where Atg12-Atg5 and LC3 II are recruited in the mature autophagosome. Lysosome fussion to the autophagosome forms the active autophagolysosome where proteins and organelles are finally degraded.

functional BH3-like domain was identified in Beclin-1 and its mutation disrupted the interaction of Beclin-1 with BCL-X$_L$.[83] Expression of BAD decreased the interaction between Beclin-1 and BCL-X$_L$ at the ER membrane and experiments performed in *C. elegans* deficient in EGL-1, a BH3 containing protein, corroborated this model in vivo. Similarly, the pharmacological BH3 mimetic ABT-737 competitively inhibited the interaction between Beclin-1 and BCL-2/BCL-X$_L$, stimulating autophagy.[83,84] Finally, a recent report indicates that phosphorylation of BCL-2 by JNK is essential for the control of autophagy by Beclin-1 (Fig 4).[85]

Along the same lines, a connection between ER calcium homeostasis and autophagy was proposed to occur through the IP3R[86-88] (reviewed in ref. 89). Stimuli that increase cytosolic calcium activate autophagy, which is blocked by BCL-2.[88] Blocking the IP3R modulates autophagy that arises from specifically inhibiting BCL-2 or BCL-X$_L$ targeting to the ER membrane. Unexpectedly, IP3R-dependent autophagy was attributed to the activation of Beclin-1 and other autophagic related genes and this regulation was independent of calcium release, possibly due to a role of IP3R as a scaffold protein rather than a calcium channel. Finally, many laboratories have now shown that ER stress triggers autophagy, effect that is also regulated by UPR stress sensors.[86,90-95] Autophagy may serve as a mechanism to eliminate damaged ER under stress conditions or to control the rate of ER expansion.[93] Unexpectedly, the activation of autophagy by ER stress requires IRE1α and is not inhibited by BCL-2 overexpression, a strategy that blocks autophagy mediated by IP3R, suggesting that there are two independent pathways controlling autophagy from the ER.[86] The direct phosphorylation of eIF2α (eukaryotic initiation factor 2) by PERK under ER stress condition is required for the LC3 conversion, a fundamental process for autophagy (Fig. 4).[90]

The BCL-2 Protein Family and ER Morphogenesis

As described above ER stress regulates the expression of BCL-2 family of proteins, interestingly some proapoptotic members of the family are also implicated in the regulation of ER structure. It has been recently shown that coexpression of BAK and BCL-X_L or BAK mutant provoke extensive swelling and vacuolization of ER cisternae.[96] Moreover, expression of upstream BH3-only activators in similar conditions recapitulates ER swelling and vacuolization when RyR activity is inhibited. This effect is mediated by BAK but not by BAX.[96] In addition, BAX and BAK may regulate the ER biogenesis through the interaction with IRE1α and activation of XBP-1.[37] Moreover, experiments in non-apoptotic cells revealed a role of BAX and BAK in mitochondrial morphogenesis. Interestingly, it seems that these two proapoptotic multidomain proteins are required for normal fusion of mitochondria into elongated tubules.[97]

The BH3-only protein BNip1 is another proapototic protein involved in the regulation of ER structure. In HeLa cells it has been shown that BNip1 is a component of a protein complex comprising syntaxin18, an ER-located soluble N-ethylmaleimide-sensitive factor (NSF) attachment protein receptor (SNARE).[98] In addition, the BH3-only domain was shown to play an important role in the binding of α-SNAP, an adaptor that serves as a link between the chaperone ATPase NSF and SNAREs. These results, together with the known apoptotic function of BNip1, unmasked a possible crosstalk between apparently independent cellular events such as apoptosis and ER membrane fusion.[98] Interestingly, protein linked to mitochondrial dynamics (fusion and fission) were also shown to modulate ER morphogenesis and fragmentation and was linked with ER-mitochondria contact sites.[99]

Conclusion

In this chapter we have summarized evidence supporting an evolutionary process whereby key regulators of cell death also contribute vital functions in cellular processes beyond apoptosis. Accordingly, BCL-2-related proteins do not only operate as upstream regulators of caspases, but they also actively regulate specific cellular functions related to ER physiology. In support of this idea, there is growing evidence of non-apoptotic functions for the BCL-2 family including cell cycle regulation,[100,101] participation in DNA damage responses (i.e., BID),[102,103] inflammation[104] and, glucose/energy metabolism (i.e., BAD).[105,106] In doing so, the BCL-2 protein family may operate as specialized stress sentinels that actively participate in crucial processes, allowing constant homeostatic quality control to respond to irreversible cellular damage activating apoptosis.

Mutations in certain genes are responsible for a variety of neurological disorders due to the misfolding and accumulation of abnormal protein aggregates in the brain. In many of these diseases alteration of ER homeostasis contributes to neuronal dysfunction. These diseases include Parkinson's disease,[107,108] Alzheimer's disease,[109] Prion diseases,[110-112] Amyotrophic Lateral Sclerosis (ALS),[113] Huntington's disease[114,115] and many others (see list of diseases in ref. 9). Consequently, the first steps in the death pathways downstream of ER stress may represent important therapeutic targets. In addition, BH3-only proteins, such as BIM, have been implicated in Alzheimer's disease[9] and ALS[116] in vivo. Thus, pharmacological manipulation of BCL-2 family members activity may be beneficial in the treatment of these fatal diseases. A number of small molecules and synthetic peptides are currently available with proven therapeutic applications in disease mouse models, including BCL-2 inhibitors,[117,118] BAX channel inhibitors,[119] BAX/BAK activator peptides[120,121] and many others (see reviews in refs. 7, 122). These drugs may be used as pharmacological tools to manipulate the activity of stress signaling pathways regulated by the BCL-2 protein family (i.e., autophagy, calcium metabolism or the UPR) and alter their function in pathological conditions.

Acknowledgements

This work was supported by FONDECYT no. 1070444, FONDAP grant no. 15010006, Millennium Nucleus no. P07-048-F, The Muscular Dystrophy Association, The Michael J. Fox Foundation for Parkinson's Research, National Parkinson's Disease, CHDI-High Q Foundation (CH), CONICYT fellowship (DR and FL) and CONICYT AT-24090143 (DR).

References

1. Ferri KF, Kroemer G. Organelle-specific initiation of cell death pathways. Nat Cell Biol 2001; 3(11):E255-63.
2. Danial NN, Korsmeyer SJ. Cell death: critical control points. Cell 2004; 116(2):205-19.
3. Strasser A. The role of BH3-only proteins in the immune system. Nat Rev Immunol 2005; 5(3):189-200.
4. Zhang HM, Cheung P, Yanagawa B et al. BNips: a group of pro-apoptotic proteins in the BCL-2 family. Apoptosis 2003, 8(3):229-236.
5. Labi V, Erlacher M, Kiessling S et al. BH3-only proteins in cell death initiation, malignant disease and anticancer therapy. Cell Death Differ 2006, 13(8):1325-1338.
6. Wei MC, Zong WX, Cheng EH et al. Proapoptotic BAX and BAK: a requisite gateway to mitochondrial dysfunction and death. Science 2001, 292(5517):727-730.
7. Reed JC. Proapoptotic multidomain BCL-2/Bax-family proteins: mechanisms, physiological roles, and therapeutic opportunities. Cell Death Differ 2006, 13(8):1378-1386.
8. Oakes SA, Lin SS, Bassik MC. The control of endoplasmic reticulum-initiated apoptosis by the BCL-2 family of proteins. Curr Mol Med 2006, 6(1):99-109.
9. Schroder M, Kaufman RJ. The mammalian unfolded protein response. Annu Rev Biochem 2005, 74:739-789.
10. Federovitch CM, Ron D, Hampton RY. The dynamic ER: experimental approaches and current questions. Curr Opin Cell Biol 2005, 17(4):409-414.
11. Rao RV, Bredesen DE. Misfolded proteins, endoplasmic reticulum stress and neurodegeneration. Curr Opin Cell Biol 2004, 16(6):653-662.
12. Koumenis C. ER stress, hypoxia tolerance and tumor progression. Curr Mol Med 2006, 6(1):55-69.
13. Lipson KL, Fonseca SG, Urano F. Endoplasmic reticulum stress-induced apoptosis and auto-immunity in diabetes. Curr Mol Med 2006, 6(1):71-77.
14. Hetz C, Glimcher L. The daily job of night killers: alternative roles of the BCL-2 family in organelle physiology. Trends Cell Biol 2008, 18(1):38-44.
15. Cox JS, Walter P. A novel mechanism for regulating activity of a transcription factor that controls the unfolded protein response. Cell 1996, 87(3):391-404.
16. Sidrauski C, Walter P. The transmembrane kinase Ire1p is a site-specific endonuclease that initiates mRNA splicing in the unfolded protein response. Cell 1997, 90(6):1031-1039.
17. Shamu CE, Walter P. Oligomerization and phosphorylation of the Ire1p kinase during intracellular signaling from the endoplasmic reticulum to the nucleus. EMBO J 1996, 15(12):3028-3039.
18. Gonzalez TN, Walter P. Ire1p: a kinase and site-specific endoribonuclease. Methods Mol Biol 2001, 160:25-36.
19. Lee K, Tirasophon W, Shen X et al. IRE1-mediated unconventional mRNA splicing and S2P-mediated ATF6 cleavage merge to regulate XBP1 in signaling the unfolded protein response. Genes Dev 2002, 16(4):452-466.
20. Calfon M, Zeng H, Urano F et al. IRE1 couples endoplasmic reticulum load to secretory capacity by processing the XBP-1 mRNA. Nature 2002, 415(6867):92-96.
21. Yoshida H, Matsui T, Yamamoto A et al. XBP1 mRNA is induced by ATF6 and spliced by IRE1 in response to ER stress to produce a highly active transcription factor. Cell 2001, 107(7):881-891.
22. Lee AH, Iwakoshi NN, Glimcher LH. XBP-1 regulates a subset of endoplasmic reticulum resident chaperone genes in the unfolded protein response. Mol Cell Biol 2003, 23(21):7448-7459.
23. Reimold AM, Etkin A, Clauss I et al. An essential role in liver development for transcription factor XBP-1. Genes Dev 2000, 14(2):152-157.
24. Reimold AM, Iwakoshi NN, Manis J et al. Plasma cell differentiation requires the transcription factor XBP-1. Nature 2001, 412(6844):300-307.
25. Iwakoshi NN, Lee AH, Vallabhajosyula P et al. Plasma cell differentiation and the unfolded protein response intersect at the transcription factor XBP-1. Nat Immunol 2003, 4(4):321-329.
26. Lee AH, Chu GC, Iwakoshi NN et al. XBP-1 is required for biogenesis of cellular secretory machinery of exocrine glands. EMBO J 2005, 24(24):4368-4380.
27. Urano F, Wang X, Bertolotti A, Zhang Y et al. Coupling of stress in the ER to activation of JNK protein kinases by transmembrane protein kinase IRE1. Science 2000, 287(5453):664-666.
28. Nishitoh H, Matsuzawa A, Tobiume K et al. ASK1 is essential for endoplasmic reticulum stress-induced neuronal cell death triggered by expanded polyglutamine repeats. Genes Dev 2002, 16(11):1345-1355.
29. Nguyen DT, Kebache S, Fazel A et al. Nck-dependent activation of extracellular signal-regulated kinase-1 and regulation of cell survival during endoplasmic reticulum stress. Mol Biol Cell 2004, 15(9):4248-4260.

30. Gu F, Nguyen DT, Stuible M et al. Protein-tyrosine phosphatase 1B potentiates IRE1 signaling during endoplasmic reticulum stress. J Biol Chem 2004, 279(48):49689-49693.
31. Hu P, Han Z, Couvillon AD et al. Autocrine tumor necrosis factor alpha links endoplasmic reticulum stress to the membrane death receptor pathway through IRE1alpha-mediated NF-kappaB activation and down-regulation of TRAF2 expression. Mol Cell Biol 2006, 26(8):3071-3084.
32. Hetz CA, Soto C. Emerging roles of the unfolded protein response signaling in physiology and disease. Curr Mol Med 2006, 6(1):1.
33. Luo D, He Y, Zhang H et al. AIP1 is critical in transducing IRE1-mediated endoplasmic reticulum stress response. J Biol Chem 2008, 283(18):11905-11912.
34. Lin JH, Li H, Yasumura D, Cohen HR et al. IRE1 signaling affects cell fate during the unfolded protein response. Science 2007, 318(5852):944-949.
35. Kuwana T, Bouchier-Hayes L, Chipuk JE et al. BH3 domains of BH3-only proteins differentially regulate Bax-mediated mitochondrial membrane permeabilization both directly and indirectly. Mol Cell 2005, 17(4):525-535.
36. Szegezdi E, Duffy A, O'Mahoney ME et al. ER stress contributes to ischemia-induced cardiomyocyte apoptosis. Biochem Biophys Res Commun 2006, 349(4):1406-1411.
37. Hetz C, Bernasconi P, Fisher J et al. Proapoptotic BAX and BAK modulate the unfolded protein response by a direct interaction with IRE1alpha. Science 2006, 312(5773):572-576.
38. Klee M, Pallauf K, Alcala S, Fleischer A et al. Mitochondrial apoptosis induced by BH3-only molecules in the exclusive presence of endoplasmic reticular Bak. EMBO J 2009.
39. Xu Q, Reed JC. Bax inhibitor-1, a mammalian apoptosis suppressor identified by functional screening in yeast. Mol Cell 1998, 1(3):337-346.
40. Chae HJ, Ke N, Kim HR, Chen S et al. Evolutionarily conserved cytoprotection provided by Bax Inhibitor-1 homologs from animals, plants, and yeast. Gene 2003, 323:101-113.
41. Huckelhoven R. BAX Inhibitor-1, an ancient cell death suppressor in animals and plants with prokaryotic relatives. Apoptosis 2004, 9(3):299-307.
42. Chae HJ, Kim HR, Xu C et al. BI-1 regulates an apoptosis pathway linked to endoplasmic reticulum stress. Mol Cell 2004, 15(3):355-366.
43. Bailly-Maitre B, Fondevila C, Kaldas F et al. Cytoprotective gene bi-1 is required for intrinsic protection from endoplasmic reticulum stress and ischemia-reperfusion injury. Proc Natl Acad Sci U S A 2006, 103(8):2809-2814.
44. Lee GH, Kim HK, Chae SW et al. Bax inhibitor-1 regulates endoplasmic reticulum stress-associated reactive oxygen species and heme oxygenase-1 expression. J Biol Chem 2007, 282(30):21618-21628.
45. Lisbona F, Rojas-Rivera D, Thielen P et al.BAX inhibitor-1 is a negative regulator of the ER stress sensor IRE1alpha. Mol Cell 2009, 33(6):679-691.
46. Kawai-Yamada M, Ohori Y, Uchimiya H. Dissection of Arabidopsis Bax inhibitor-1 suppressing Bax-, hydrogen peroxide-, and salicylic acid-induced cell death. Plant Cell 2004, 16(1):21-32.
47. Hetz C, Bono MR, Barros LF, Lagos R. Microcin E492, a channel-forming bacteriocin from Klebsiella pneumoniae, induces apoptosis in some human cell lines. Proc Natl Acad Sci U S A 2002, 99(5):2696-2701.
48. Kim H, Rafiuddin-Shah M, Tu HC et al. Hierarchical regulation of mitochondrion-dependent apoptosis by BCL-2 subfamilies. Nat Cell Biol 2006, 8(12):1348-1358.
49. Willis SN, Fletcher JI, Kaufmann T et al. Apoptosis initiated when BH3 ligands engage multiple BCL-2 homologs, not Bax or Bak. Science 2007, 315(5813):856-859.
50. Letai A, Bassik MC, Walensky LD et al. Distinct BH3 domains either sensitize or activate mitochondrial apoptosis, serving as prototype cancer therapeutics. Cancer Cell 2002, 2(3):183-192.
51. Li J, Lee B, Lee AS. Endoplasmic reticulum stress-induced apoptosis: multiple pathways and activation of p53-up-regulated modulator of apoptosis (PUMA) and NOXA by p53. J Biol Chem 2006, 281(11):7260-7270.
52. Reimertz C, Kogel D, Rami A et al. Gene expression during ER stress-induced apoptosis in neurons: induction of the BH3-only protein Bbc3/PUMA and activation of the mitochondrial apoptosis pathway. J Cell Biol 2003, 162(4):587-597.
53. Futami T, Miyagishi M, Taira K. Identification of a network involved in thapsigargin-induced apoptosis using a library of small interfering RNA expression vectors. J Biol Chem 2005, 280(1):826-831.
54. Jiang CC, Lucas K, Avery-Kiejda KA et al. Up-regulation of Mcl-1 is critical for survival of human melanoma cells upon endoplasmic reticulum stress. Cancer Res 2008, 68(16):6708-6717.
55. Wang Q, Mora-Jensen H, Weniger MAet al. ERAD inhibitors integrate ER stress with an epigenetic mechanism to activate BH3-only protein NOXA in cancer cells. Proc Natl Acad Sci U S A 2009, 106(7):2200-2205.
56. Germain M, Mathai JP, Shore GC. BH-3-only BIK functions at the endoplasmic reticulum to stimulate cytochrome c release from mitochondria. J Biol Chem 2002, 277(20):18053-18060.

57. Mathai JP, Germain M, Shore GC. BH3-only BIK regulates BAX,BAK-dependent release of Ca2+ from endoplasmic reticulum stores and mitochondrial apoptosis during stress-induced cell death. J Biol Chem 2005, 280(25):23829-23836.
58. Puthalakath H, O'Reilly LA, Gunn P et al. ER stress triggers apoptosis by activating BH3-only protein Bim. Cell 2007, 129(7):1337-1349.
59. Morishima N, Nakanishi K, Tsuchiya K et al. Translocation of Bim to the endoplasmic reticulum (ER) mediates ER stress signaling for activation of caspase-12 during ER stress-induced apoptosis. J Biol Chem 2004, 279(48):50375-50381.
60. Elyaman W, Terro F, Suen KC et al. BAD and BCL-2 regulation are early events linking neuronal endoplasmic reticulum stress to mitochondria-mediated apoptosis. Brain Res Mol Brain Res 2002, 109(1-2):233-238.
61. Upton JP, Austgen K, Nishino M et al. Caspase-2 cleavage of BID is a critical apoptotic signal downstream of endoplasmic reticulum stress. Mol Cell Biol 2008, 28(12):3943-3951.
62. Scorrano L, Oakes SA, Opferman JT et al. BAX and BAK regulation of endoplasmic reticulum Ca2+: a control point for apoptosis. Science 2003, 300(5616):135-139.
63. Zong WX, Li C, Hatzivassiliou G et al. Bax and Bak can localize to the endoplasmic reticulum to initiate apoptosis. J Cell Biol 2003, 162(1):59-69.
64. Bassik MC, Scorrano L, Oakes SA et al. Phosphorylation of BCL-2 regulates ER Ca2+ homeostasis and apoptosis. EMBO J 2004, 23(5):1207-1216.
65. Breckenridge DG, Germain M, Mathai JP et al. Regulation of apoptosis by endoplasmic reticulum pathways. Oncogene 2003, 22(53):8608-8618.
66. Pinton P, Rizzuto R. BCL-2 and Ca2+ homeostasis in the endoplasmic reticulum. Cell Death Differ 2006, 13(8):1409-1418.
67. Xu C, Xu W, Palmer AE, Reed JC. BI-1 regulates endoplasmic reticulum Ca2+ homeostasis downstream of BCL-2 family proteins. J Biol Chem 2008, 283(17):11477-11484.
68. Westphalen BC, Wessig J, Leypoldt F et al. BI-1 protects cells from oxygen glucose deprivation by reducing the calcium content of the endoplasmic reticulum. Cell Death Differ 2005, 12(3):304-306.
69. Oakes SA, Scorrano L, Opferman JT et al. Proapoptotic BAX and BAK regulate the type 1 inositol trisphosphate receptor and calcium leak from the endoplasmic reticulum. Proc Natl Acad Sci U S A 2005, 102(1):105-110.
70. Chen R, Valencia I, Zhong F, McColl KS et al.. BCL-2 functionally interacts with inositol 1,4,5-trisphosphate receptors to regulate calcium release from the ER in response to inositol 1,4,5-trisphosphate. J Cell Biol 2004, 166(2):193-203.
71. White C, Li C, Yang J, Petrenko NB et al. The endoplasmic reticulum gateway to apoptosis by Bcl-X(L) modulation of the InsP3R. Nat Cell Biol 2005, 7(10):1021-1028.
72. Lin SS, Bassik MC, Suh H et al. PP2A regulates BCL-2 phosphorylation and proteasome-mediated degradation at the endoplasmic reticulum. J Biol Chem 2006, 281(32):23003-23012.
73. Dremina ES, Sharov VS, Kumar K et al. Anti-apoptotic protein BCL-2 interacts with and destabilizes the sarcoplasmic/endoplasmic reticulum Ca2+-ATPase (SERCA). Biochem J 2004, 383(Pt 2):361-370.
74. Dremina ES, Sharov VS, Schoneich C. Displacement of SERCA from SR lipid caveolae-related domains by BCL-2: a possible mechanism for SERCA inactivation. Biochemistry 2006, 45(1):175-184.
75. Levine B. Eating oneself and uninvited guests: autophagy-related pathways in cellular defense. Cell 2005, 120(2):159-162.
76. Cuervo AM. Autophagy: in sickness and in health. Trends Cell Biol 2004, 14(2):70-77.
77. Aita VM, Liang XH, Murty VV et al. Cloning and genomic organization of beclin 1, a candidate tumor suppressor gene on chromosome 17q21. Genomics 1999, 59(1):59-65.
78. Liang XH, Jackson S, Seaman M et al. Induction of autophagy and inhibition of tumorigenesis by beclin 1. Nature 1999, 402(6762):672-676.
79. Qu X, Yu J, Bhagat G, Furuya N et al. Promotion of tumorigenesis by heterozygous disruption of the beclin 1 autophagy gene. J Clin Invest 2003, 112(12):1809-1820.
80. Yue Z, Jin S, Yang C, Levine AJ et al. Beclin 1, an autophagy gene essential for early embryonic development, is a haploinsufficient tumor suppressor. Proc Natl Acad Sci U S A 2003, 100(25):15077-15082.
81. Pattingre S, Tassa A, Qu X et al. BCL-2 antiapoptotic proteins inhibit Beclin 1-dependent autophagy. Cell 2005, 122(6):927-939.
82. Ciechomska IA, Goemans GC, Skepper JN et al. BCL-2 complexed with Beclin-1 maintains full anti-apoptotic function. Oncogene 2009, 28(21):2128-2141.
83. Maiuri MC, Le Toumelin G, Criollo A et al. Functional and physical interaction between Bcl-X(L) and a BH3-like domain in Beclin-1. EMBO J 2007, 26(10):2527-2539.
84. Maiuri MC, Criollo A, Tasdemir E et al. BH3-only proteins and BH3 mimetics induce autophagy by competitively disrupting the interaction between Beclin 1 and BCL-2/Bcl-X(L). Autophagy 2007, 3(4):374-376.

85. Wei Y, Pattingre S, Sinha S et al. JNK1-mediated phosphorylation of BCL-2 regulates starvation-induced autophagy. Mol Cell 2008, 30(6):678-688.
86. Criollo A, Vicencio JM, Tasdemir E et al. The inositol trisphosphate receptor in the control of autophagy. Autophagy 2007, 3(4):350-353.
87. Criollo A, Maiuri MC, Tasdemir E et al. Regulation of autophagy by the inositol trisphosphate receptor. Cell Death Differ 2007, 14(5):1029-1039.
88. Hoyer-Hansen M, Bastholm L, Szyniarowski P et al. Control of macroautophagy by calcium, calmodulin-dependent kinase kinase-beta, and BCL-2. Mol Cell 2007, 25(2):193-205.
89. Hoyer-Hansen M, Jaattela M. Connecting endoplasmic reticulum stress to autophagy by unfolded protein response and calcium. Cell Death Differ 2007, 14(9):1576-1582.
90. Kouroku Y, Fujita E, Tanida I et al. ER stress (PERK/eIF2alpha phosphorylation) mediates the polyglutamine-induced LC3 conversion, an essential step for autophagy formation. Cell Death Differ 2007, 14(2):230-239.
91. Ogata M, Hino S, Saito A et al. Autophagy is activated for cell survival after endoplasmic reticulum stress. Mol Cell Biol 2006, 26(24):9220-9231.
92. Yorimitsu T, Nair U, Yang Z et al. Endoplasmic reticulum stress triggers autophagy. J Biol Chem 2006, 281(40):30299-30304.
93. Bernales S, McDonald KL, Walter P. Autophagy counterbalances endoplasmic reticulum expansion during the unfolded protein response. PLoS Biol 2006, 4(12):e423.
94. Ding WX, Ni HM, Gao W et al. Differential effects of endoplasmic reticulum stress-induced autophagy on cell survival. J Biol Chem 2007, 282(7):4702-4710.
95. Ding WX, Ni HM, Gao W et al. Linking of autophagy to ubiquitin-proteasome system is important for the regulation of endoplasmic reticulum stress and cell viability. Am J Pathol 2007, 171(2):513-524.
96. Klee M, Pimentel-Muinos FX. Bcl-X(L) specifically activates Bak to induce swelling and restructuring of the endoplasmic reticulum. J Cell Biol 2005, 168(5):723-734.
97. Karbowski M, Norris KL, Cleland MM et al. Role of Bax and Bak in mitochondrial morphogenesis. Nature 2006, 443(7112):658-662.
98. Nakajima K, Hirose H, Taniguchi M et al. Involvement of BNIP1 in apoptosis and endoplasmic reticulum membrane fusion. EMBO J 2004, 23(16):3216-3226.
99. de Brito OM, Scorrano L. Mitofusin 2 tethers endoplasmic reticulum to mitochondria. Nature 2008, 456(7222):605-610.
100. Schuler M, Green DR. Transcription, apoptosis and p53: catch-22. Trends Genet 2005, 21(3):182-187.
101. Zinkel S, Gross A, Yang E. BCL2 family in DNA damage and cell cycle control. Cell Death Differ 2006, 13(8):1351-1359.
102. Kamer I, Sarig R, Zaltsman Y et al. Proapoptotic BID is an ATM effector in the DNA-damage response. Cell 2005, 122(4):593-603.
103. Zinkel SS, Hurov KE, Ong C et al. A role for proapoptotic BID in the DNA-damage response. Cell 2005, 122(4):579-591.
104. Bruey JM, Bruey-Sedano N, Luciano F et al. BCL-2 and BCL-XL regulate proinflammatory caspase-1 activation by interaction with NALP1. Cell 2007, 129(1):45-56.
105. Danial NN, Gramm CF, Scorrano L et al. BAD and glucokinase reside in a mitochondrial complex that integrates glycolysis and apoptosis. Nature 2003, 424(6951):952-956.
106. Danial NN, Walensky LD, Zhang CY et al. Dual role of proapoptotic BAD in insulin secretion and beta cell survival. Nat Med 2008, 14(2):144-153.
107. Holtz WA, O'Malley KL. Parkinsonian mimetics induce aspects of unfolded protein response in death of dopaminergic neurons. J Biol Chem 2003, 278(21):19367-19377.
108. Ryu EJ, Harding HP, Angelastro JM et al. Endoplasmic reticulum stress and the unfolded protein response in cellular models of Parkinson's disease. J Neurosci 2002, 22(24):10690-10698.
109. Ghribi O. The role of the endoplasmic reticulum in the accumulation of beta-amyloid peptide in Alzheimer's disease. Curr Mol Med 2006, 6(1):119-133.
110. Hetz C, Russelakis-Carneiro M, Maundrell K et al. Caspase-12 and endoplasmic reticulum stress mediate neurotoxicity of pathological prion protein. EMBO J 2003, 22(20):5435-5445.
111. Hetz C, Russelakis-Carneiro M, Walchli S et al. The disulfide isomerase Grp58 is a protective factor against prion neurotoxicity. J Neurosci 2005, 25(11):2793-2802.
112. Hetz CA, Soto C. Stressing out the ER: a role of the unfolded protein response in prion-related disorders. Curr Mol Med 2006, 6(1):37-43.
113. Turner BJ, Atkin JD. ER stress and UPR in familial amyotrophic lateral sclerosis. Curr Mol Med 2006, 6(1):79-86.
114. Momoi T. Conformational diseases and ER stress-mediated cell death: apoptotic cell death and autophagic cell death. Curr Mol Med 2006, 6(1):111-118.

115. Sekine Y, Takeda K, Ichijo H. The ASK1-MAP kinase signaling in ER stress and neurodegenerative diseases. Curr Mol Med 2006, 6(1):87-97.
116. Hetz C, Thielen P, Fisher J et al. The proapoptotic BCL-2 family member BIM mediates motoneuron loss in a model of amyotrophic lateral sclerosis. Cell Death Differ 2007, 14(7):1386-1389.
117. Oltersdorf T, Elmore SW, Shoemaker AR et al. An inhibitor of BCL-2 family proteins induces regression of solid tumours. Nature 2005, 435(7042):677-681.
118. Nguyen M, Marcellus RC, Roulston A et al. Small molecule obatoclax (GX15-070) antagonizes MCL-1 and overcomes MCL-1-mediated resistance to apoptosis. Proc Natl Acad Sci U S A 2007, 104(49):19512-19517.
119. Hetz C, Vitte PA, Bombrun A, Rostovtseva TK et al. Bax channel inhibitors prevent mitochondrion-mediated apoptosis and protect neurons in a model of global brain ischemia. J Biol Chem 2005, 280(52):42960-42970.
120. Walensky LD, Kung AL, Escher I et al. Activation of apoptosis in vivo by a hydrocarbon-stapled BH3 helix. Science 2004, 305(5689):1466-1470.
121. Walensky LD, Pitter K, Morash J et al. A stapled BID BH3 helix directly binds and activates BAX. Mol Cell 2006, 24(2):199-210.
122. Letai A. Pharmacological manipulation of BCL-2 family members to control cell death. J Clin Invest 2005, 115(10):2648-2655.

CHAPTER 3

BH3-Only Proteins and Their Effects on Cancer

Thanh-Trang Vo and Anthony Letai*

Abstract

Apoptosis, a form of cellular suicide is a key mechanism involved in the clearance of cells that are dysfunctional, superfluous or infected. For this reason, the cell needs mechanisms to sense death cues and relay death signals to the apoptotic machinery involved in cellular execution. In the intrinsic apoptotic pathway, a subclass of BCL-2 family proteins called the BH3-only proteins are responsible for triggering apoptosis in response to varied cellular stress cues. The mechanisms by which they are regulated are tied to the type of cellular stress they sense. Once triggered, they interact with other BCL-2 family proteins to cause mitochondrial outer membrane permeabilization which in turn results in the activation of serine proteases necessary for cell killing. Failure to properly sense death cues and relay the death signal can have a major impact on cancer. This chapter will discuss our current models of how BH3-only proteins function as well as their impact on carcinogenesis and cancer treatment.

Introduction

In multicellular organisms, various cells of differing specialized functions work together to ensure the survival of the whole organism. In this complex cooperative network, it is essential to ensure the clearance of dysfunctional cells that may pose a risk to the collective. Thus the machinery required to carry out the cellular suicide program known as apoptosis is programmed genetically into each cell. Apoptosis is a form of programmed cell death that is essential in the clearance of cells that are infected, dislocated from their normal positions, damaged, superfluous or have reached the end of their useful life span. Once the apoptotic pathway is engaged, cells are efficiently dismantled and cleared. This efficient process is mediated by the activation of caspases, which are a family of specialized serine proteases that effectively cleave various protein substrates within the cell. One result of their proteolytic activity is the activation of the endonuclease CAD (caspase-activated DNase) which goes on to dismantle the cellular genome, preventing replication of the undesirable clone.[1] The dying cells also exhibit cell surface markers that flag them for engulfment and clearance by macrophages. Once caspases have cleaved their downstream substrates, the destruction is irreparable and cell death is inescapable. For this reason, the pathways that lead to caspase activation are critical in determining cell fate.

Apoptosis can be executed by two major pathways called the extrinsic and intrinsic pathways. In the extrinsic pathway, extracellular death signals in the form of ligands bind and activate cell membrane-anchored death receptors like FAS (also known as CD95) receptor, TNF (tumor necrosis factor) receptor and TRAIL (TNF-related apoptosis-inducing ligand) receptor.[2] After ligand binding, death receptors aggregate and recruit the adaptor molecule FADD (Fas

*Corresponding Author: Anthony Letai—Dana-Farber Cancer Institute, Harvard Medical School, Boston, Massachusetts 02115, USA. Email: anthony_letai@dfci.harvard.edu

BCL-2 Protein Family: Essential Regulators of Cell Death, edited by Claudio Hetz.
©2010 Landes Bioscience and Springer Science+Business Media.

Associated Death Domain). FADD interacts with pro-caspase-8 to form a complex known as the Death Inducing Signaling Complex (DISC). This complex places several pro-caspase-8 proteins in proximity to each other, causing them to activate by cleavage.[2] Fully activated caspase-8 is an initiator caspase that goes on to cleave and activate effector caspases needed to kill the cell.

The intrinsic pathway relies on the mitochondria and thus is also referred to as the mitochondrial pathway. In this pathway, the cell internally senses death cues and usually relays the death signal through a subclass of BCL-2 family proteins called the BH3-only members. These BH3-only proteins interact with other pro-apoptotic and anti-apoptotic proteins members of the BCL-2 family to decide cell fate. If the cell commits to death, pro-apoptotic BCL-2 family members cause the mitochondrial outer membrane to become permeabilized (MOMP) and apoptogenic factors like cytochrome c are released into the cytosol.[3-5] Once in the cytosol, cytochrome c interacts with APAF-1 (apoptotic protease activating factor 1) and pro-caspase-9 to form a complex termed the apoptosome.[6] The apoptosome complex facilitates the proximity induced auto-cleave of pro-caspase-9 to the active caspase-9.[7,8] Similar to caspase-8, caspase-9 is an initiator caspase that goes on to cleave and activate other effector caspases to kill the cell.

In many cells, activation of the extrinsic pathway alone is insufficient to induce apoptosis.[2] Instead, recruitment of the intrinsic pathway is also required. Caspase-8 can amplify the extrinsic death signal by cleaving and activating the BH3-only protein Bid to trigger activation of the intrinsic pathway. Thus, BH3-only proteins are important players responsible for communicating death signals originating in both the extrinsic and intrinsic pathways. Understanding how these BH3-only proteins function will help us not only understand how cells survive to become cancerous but also how to trigger these death cues for better chemotherapeutics.

Categories of BCL-2 Family Proteins and Their Apoptotic Functions

The BCL-2 family of proteins consists of three main categories based on their function and sequence homology as shown on Figure 1. Each member of this family of proteins shares at least one of four regions of homology with their founding member BCL-2. These BCL-2 homology regions are often denoted as BH1 through BH4. The multidomain anti-apoptotic members share all four regions (except Mcl-1 and Bfl-1). The multidomain pro-apoptotic members share domains BH1, BH2 and BH3. The final group consists of pro-apoptotic proteins that only share the BH3 domain and are thus referred to as the BH3-only proteins.

Multidomain Proapoptotic Members

The multidomain proapoptotic BCL-2 family proteins consist of Bax, Bak and Bok. Bax and Bak are prevalent and expressed in a wide variety of tissues, whereas the less studied Bok is only known to be prevalent in reproductive tissue. Bax and Bak proteins consist of five amphipathic α-helices surrounding two central hydrophobic α-helices long enough to traverse a lipid bilayer.[9] These proteins form pores in the outer membrane of the mitochondria in response to death signals. The resulting mitochondrial outer membrane permeability (MOMP) causes the release of death factors like cytochrome c into the cytosol to execute cell killing. Indeed, Bax was found to permeabilize giant liposomes releasing their fluorescence dye content.[10] Furthermore, cytochrome c release is not observed in Bax/Bak null cells during drug treatments that normally evoke apoptosis.[11] BH3-only proteins lack pro-death function in the absence of Bax and Bak.[11] Thus, the roles of Bax and Bak in executing MOMP makes them essential for apoptosis to occur through the intrinsic apoptotic pathway.[11]

In the absence of a death signal, Bax and Bak mainly exist as monomers in their inactive forms. Inactive Bax can be found in the cytosol or loosely associated with the outer mitochondrial membrane. Inactive Bak is inserted into the mitochondrial outer membrane.[12] In the presence of a death signal, Bax and Bak become activated and undergo a conformational change that exposes the N-terminal of the proteins. Upon activation, Bax translocates and inserts into the mitochondria. Both activated Bax and Bak homo-oligomerize and form pores to cause MOMP. As detailed below, the steps that lead to Bax and Bak induction of MOMP can be regulated by the other BCL-2

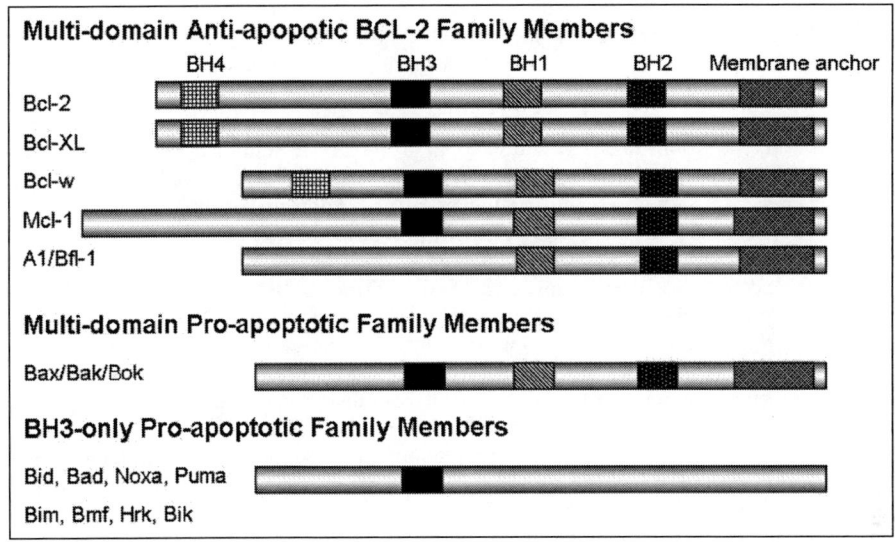

Figure 1. BCL-2 homology domains of BCL-2 family members.

family proteins. Both the anti-apoptotic and BH3-only category of proteins have important roles in regulating Bax and Bak activity.

Anti-Apoptotic BCL-2 Family Members

The multidomain anti-apoptotic proteins include BCL-2, BCL-XL, BCL-2, Mcl-1 and Bfl-1/A1. BCL-2 and its two closest homologs BCL-XL and BCL-2 share the same core structure as the pore forming Bax and Bak, which consist of five amphipathic α-helices surrounding two central hydrophobic α-helices. These proteins possess three or four BH domains and a C-terminal transmembrane anchor. A hydrophobic groove formed by the BH1, BH2 and BH3 domains provide a binding site for the hydrophobic face of the amphipathic BH3 domain of pro-apoptotic proteins.

These anti-apoptotic family members have been observed to bind Bax and Bak resulting in the prevention of Bax/Bak oligomerization and MOMP. Modification of the hydrophobic pocket of BCL-XL inhibits this Bax interaction.[13] This interaction is greatly increased in the presence of detergents, like Triton X100 and NP-40, that induce the active conformation of Bax.[14] This suggests that anti-apoptotic BCL-2 family proteins preferentially inhibit the activated form of Bax and Bak. The BH3-domain of BH3-only proteins can also bind to the hydrophobic pockets of anti-apoptotic family members.[15] Indeed, this may be the most important anti-apoptotic function of anti-apoptotic proteins. Two main line of evidence support this view. First, a mutant BCL-XL protein that lacked the ability to bind Bax or Bak, but could still bind BH3 only proteins, maintained nearly all of its anti-apoptotic function.[16] Second, in living cells, when BCL-2 protects against cell death, it does so usually by preventing activation of Bax and Bak, suggesting that interception of Bax and Bak activating signals is the primary function of BCL-2 and related anti-apoptotic proteins.[17] Nonetheless, cells may be found in which a portion of the total Bax and Bak is indeed bound to anti-apoptotic proteins. The significance of anti-apoptotic proteins binding to BH3-only proteins in regulating Bax and Bak activity is discussed further in the next section.

Proapoptotic BH3-Only Function

Proteins which can most confidently be referred to as BH3-only proteins include Bim,[18] Bid,[19] Bad,[20] Bik,[21] Noxa,[22] Puma,[23] Hrk,[24] Bmf,[25] Mule,[26] Bcl-g.[27] Others have been identified as possessing BH3-domains, but their function as pro-death molecules in the BCL-2 family remains less clear. These include Bnip3,[28] Beclin-1,[29] ApoL6,[30] BRCC2,[31] Spike[32] and MAP-1.[33] It is worth noting that the roughly 20 amino acid BH3 domain has only a few amino acids that are highly conserved across all members, so that searching for BH3 domains by sequence alone is challenging, if not impossible. This chapter will focus on the better studied BH3-only proteins Bim, Bid, Bad, Bik, Noxa, Puma, Hrk and Bmf. These proteins are induced by cell death cues and relay their death signal to the mitochondria. As their name denotes, BH3-only proteins share homology with BCL-2 only in their BH3 domain. The BH3 domain consists of an amphipathic α-helix. Mutating the BH3-domain of BH3-only proteins abrogates their pro-apoptotic function.[18,19] Furthermore, short polypeptide oligomers made from the BH3 domain sequence are sufficient to induce Bax- and Bak-mediated MOMP.[17,34] Thus the BH3 domain of the BH3-only proteins are both necessary and sufficient for their apoptotic function.

BH3-only proteins can be divided into two different groups based on their function, the activators and the sensitizers.[34] BH3-only proteins require the presence of Bax or Bak proteins to perform their pro-apoptotic function.[35] Data suggests that certain BH3-only proteins, called activators, may directly interact with Bax and Bak to promote their activation. These activator BH3-only proteins include Bim and the truncated form of Bid (tBid). Bid was initially cloned via interaction with both Bax and BCL-2, so it is perhaps not surprising that an interaction with Bax has biological relevance.[19] Both tBid as well as the BH3 domain from Bid have been shown to induce Bax activation and membrane permeabilization in mitochondrial and liposomal systems.[12,34,36] These observations have been greatly bolstered by more recent work directly observing interaction in a lipid membrane using full-length proteins and fluorescence resonance energy transfer (FRET).[37] Also, structural studies show that the BH3 domain of Bim binds the α1 and α6 helices of Bax and not the hydrophobic pocket formed by the BH1, BH1 and BH3 domains.[38] However, since loss of both Bim and Bid has only minor effects on select triggers of apoptosis,[39] it seems likely that these factors are not the only activators. Indeed, activator function has also been attributed to p53 and the BH3-only protein Puma.[40,41] Other uncharacterized proteins may play this role and even heat has been shown to foster Bax activation.[42] Anti-apoptotic BCL-2 family proteins can bind and sequester activator BH3-only proteins, preventing them from activating Bax and Bak.[34,43]

Sensitizer BH3-only proteins lack the ability to directly activate Bax and Bak but are still pro-apoptotic. They instead bind anti-apoptotic family proteins like BCL-2 and Mcl-1 to inhibit their function.[18-22] The hydrophobic face of the BH3 α-helix inserts into the hydrophobic cleft formed by the BH1, BH2 and BH3 domains of the anti-apoptotic family members. By

	Activators		Sensitizers					
	Bid	Bim	Bad	Bik	Noxa	Puma	Hrk	Bmf
Bcl-2	white	white	white	grey	black	white	black	white
Bcl-XL	white	white	white	white	black	white	grey	white
Bcl-w	white	white	white	white	black	white	black	white
Mcl-1	white	white	black	grey	white	white	black	white
Bfl-1/A1	white	white	black	white	white	white	black	white

Figure 2. Binding specificity of various BH3-only protein oligomers for anti-apoptotic members. White blocks denotes strong affinity, grey blocks denote weak-affinity and black blocks denote no detectable binding. Adapted from: Certo M et al. Cancer Cell 2006; 9(5):351-365;[17] ©2006 with permission from Elsevier.

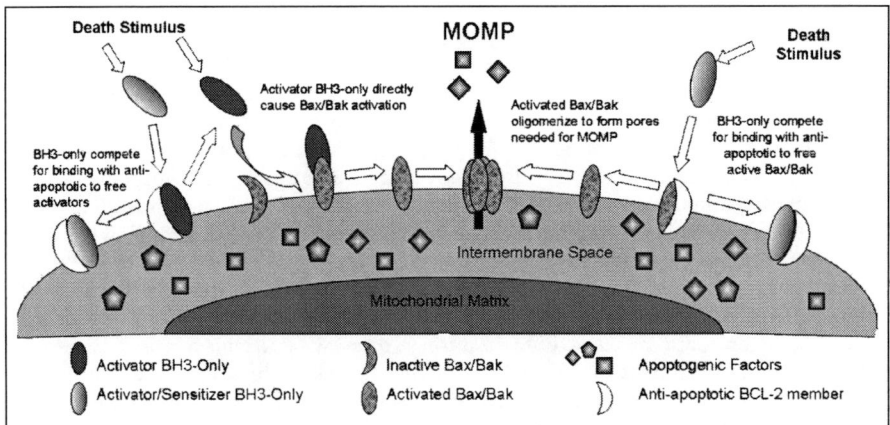

Figure 3. Model of BH3-only induction of Bax/Bak mediated outer mitochondrial membrane permeabilization.

binding to anti-apoptotic members, BH3-only proteins prevent critical anti-apoptotic binding of pro-apoptotic molecules including activators like Bim and Bid, as well as activated, monomeric Bax and Bak. If the anti-apoptotic proteins are already binding select pro-death proteins such as activator BH3-only proteins (e.g., Bim or Bid) or activated Bax or Bak, the inhibition of anti-apoptotic function can induce death by freeing those molecules to trigger MOMP. It must be noted that not all BH3-only proteins can bind all anti-apoptotic members. Binding assays show BH3 peptides from the various BH3-only proteins have specificity for only certain anti-apoptotic BCL-2 family members.[17,44] Some BH3-only proteins like Bim, Bid and Puma can bind all known anti-apoptotic family members. However, proteins like Bad, Bik, Noxa, Hrk and Bmf are more limited in the spectrum of anti-apoptotic proteins they can bind. Figure 2 displays the binding specificities of the BH3 peptide from each BH3-only protein to the anti-apoptotic family members.

The presence of two distinct models of control over apoptosis, the so-called direct and indirect models, have received considerable attention in the field. However, with the now widespread appreciation that Bax and Bak require some kind of activating step to effect MOMP, the differences are quite slender. Proponents of the direct model have focused on the ability of activator BH3-only proteins to cause death by activating Bax and Bak. Proponents of the indirect pathway have focused more on the sensitizer function of BH3-only proteins, particularly in their ability to displace Bax and Bak that is already activated. Both mechanisms likely occur in cells, the relative degree no doubt varying according to context. A summary can be found in Figure 3.

Regulation of Various BH3-Only Proteins and Cell Stress Sensing

Genetic and biochemical data has shown that, depending on context, different BH3-only proteins are essential for cell death initiation through the intrinsic pathway. Each BH3-only protein member can be stimulated by many mechanisms that correspond to the type of cell stress the protein is able to sense. However, it must be also be noted that many stimuli activate multiple BH3-only proteins to ensure the death signal is carried forth.

Bim (BCL-2 interacting mediator of cell death)[18] is constitutively expressed in several tissues and exists as a number of splice variants denoted as BimEL (the largest), BimL and Bims (the smallest). Bims has the most potent apoptosis-inducing activity.[18,45] There are other splice isoforms that are not observed as often like α1-1, β1-4 and Bim AD[46] and others.[47] However, BimEL and BimL seem to be the most relevant in vivo. Bim is regulated by multiple known mechanisms. Bim can be localized to the microtubules by binding to dynein light chain (DLC)

1, a component of the microtubule dynein motor complex.[48] This localization suggests that Bim may be activated during microtubule perturbation. Bim can also be regulated transcriptionally. After cytokine withdrawal, activation of Foxo3a (forkhead transcription factor) induces Bim expression in lymphocytes.[49] Loss of cellular adhesion can also cause transcriptional upregulation of Bim via the Mek/Erk pathway.[50] Phosphorylation by Jun N-terminal kinase (JNK) in the DLC-binding domain causes Bim to dissociate from the microtubule and induce cell death after UV irradiation.[51] Bim can also be downregulated by proteasomal degradation after ERK-mediated phosphorylation and subsequent ubiquitination.[52] With all of these mechanisms of regulation, Bim can serve as a sentinel for genotoxic stress, microtubule stress, loss of cellular adhesion and loss of growth factors.

Bid (BH3-interacting domain) is relatively inactive until proteolytically cleaved to the truncated form tBid. Bid can be cleaved by caspase-8 and thus serve as a means by which the extrinsic apoptotic pathway can activate the intrinsic pathway. Granzyme B, a protease released in secretory granules by cytotoxic T-lymphocytes and natural killer cells, can also activate Bid.[53,54] These granules are often directed at virally transfected or transformed cells. The lysosomal protein cathepsin can also cleave Bid after lysosomal disruption.[55] Once cleaved, Bid becomes myristoylated which promotes its targeting to the mitochondria.[56] Aside from proteolytic activation, Bid was also shown to be transcriptionally induced by p53 during γ-irradiation though much of it remains inactive and uncleaved.[57] Since upregulation of total Bid was not accompanied with increased activated tBid, it is not clear how this mode of Bid activation contributes to p53-induced apoptosis. Data also suggests that Bid may be regulated by ATM kinase mediated phosphorylation during the DNA damage response.[58] Nonphosphorylatable Bid mutant was more resistant to etoposide-induced apoptosis than wild-type.[58] Thus Bid is a sentinel that detects internal death cues from DNA damage as well as relay external death signals.

Bad is best known for its role in apoptosis induced by cytokine or growth factor deprivation. The Bad gene results in two different size proteins (Bad long and short) due to alternate translation initiation and translation start sites.[59] It is tightly regulated by phosphorylation through various pathways. In mice, serine 112, serine 136 and serine 155 are phosphorylated. In humans the residues phosphorylated are Ser-75, Ser-99 and Ser-118. Protein kinases that phosphorylate Bad include Akt,[60,61] Raf-1,[62] protein kinase A (PKA).[63] All of these kinases are activated by growth survival signals. Phosphorylation at these sites promotes sequestration of Bad in the cytoplasm by 14-3-3 scaffold proteins away from the mitochondria.[64] Phosphatases that dephosphorylate Bad include calcineurin,[65] protein phosphatase 1 (PP1)[66] and protein phosphatase 2A (PP2A).[67] Dephosphorylation allows Bad to translocate to the mitochondria and inhibit select anti-apoptotic BCL-2 members.

Bik (also known as Nbk) is induced due to various apoptotic triggers. The mouse orthologue is called Blk. Bik expression can be upregulated by p53.[68] Once induced, Bik translocates to the mitochondria and endoplasmic reticulum. Bik can be enhanced when phosphorylated possibly by a casein kinase II related enzyme.[69] Nonphosphorylatable Bik displays less apoptotic activity though its ability to bind BCL-2 is maintained.[69] Bik can also be transcriptionally upregulated by the oncogene E2F in a manor that is independent of p53.[70]

Puma (p53 upregulated modulator of apoptosis) was first discovered as a gene induced by p53.[23] It was concurrently discovered by another lab which named it bbc3.[71] The promoter contains two p53 binding sites allowing it to be directly affected by p53. The gene produces two proteins products (Puma-α and Puma-β) that exhibit similar activity in binding BCL-2 and causing cytochrome c release.[23] Puma substantially contributes to p53-dependent and independent apoptosis during glucocorticoid treatment, ionizing irradiation, deregulated c-Myc induced death and cell death induction during cytokine withdrawal.[71,72] Though other BCL-2 family members are also transcriptionally upregulated by p53, data suggest that Puma plays a predominant role in mediating p53-induced apoptosis.[72,73]

Noxa (Greek for "damage") was the first BH3-only protein reported to be a p53 transcriptional target.[22] The human Noxa protein has one BH3 domain but the mouse Noxa is the only

BCL-2 family member known to have two functional BH3 domains.[22] Both human and mouse Noxa genes have a p53-binding site in their promoter causing them to be a direct effector of p53.[22] Together with Puma, Noxa plays a role in p53-mediated death but seems to be less important than Puma in many cases. This may be due it is effects being more limited than Puma in regards to the anti-apoptotic BCL-2 members it can neutralize(see Figure 2). Noxa is thought to induce apoptosis though antagonism MCL-1 and BFL-1.[17,44]

Bmf is reported to be regulated in a similar manner as Bim and also has a conserved dynein light chain-binding motif. However, instead of binding to the microtubules, BMF is bound to dynein light chain 2 (DLC2) which is a component of the myosin C actin motor complex found on the actin microfilament.[25] Loss of cellular adhesion can mobilize Bmf to exert its pro-apoptotic function.[74]

Human Hrk has a murine homologue DP5, which was first discovered to induce death when rat sympathetic neurons were deprived of nerve growth factor (NGF).[75] Murine DP5 is mainly only expressed in the nervous system[75] while human Hrk is also found in hematopoietic tissues.[24] Hrk can be regulated at the transcription level. It has been shown that E2F1 can transcriptionally upregulate Hrk. Hrk is also induced during growth factor withdrawal in haemopoietic progenitor cells lines.[76] Thus, Hrk can serve as a sentinel for growth factor withdrawal and oncogene activation induced death.

Though much information is known regarding the regulation of BH3-only proteins, a full picture of what triggers cause each protein to be activated is far from complete. Also, each tissue type may utilize a different set of BH3-only to respond to the same death stimulus. Lastly, it must be noted that cancer cells exhibit at baseline many characteristics that would normally mobilize these BH3-only sentinels. Oncogenes like Myc upregulates Bim and E2F1 upregulates Hrk. Metastatic cancer cells also loose normal adhesion to their environment which should trigger Bim and Bmf. Tumors that grow outside of their normal tissue also lack normal survival factors that could result in induction of BH3-only proteins like Bim, Bad, Puma and Hrk. Thus, BH3-only proteins may play important roles in cancer prevention at very early stages of tumorigensis. In the next section, the possible impact of BH3-only proteins to cancer development will be assessed.

BH3-Only Proteins and Cancer Development

It has been postulated that inhibiting apoptosis is a prerequisite for cancer development.[77] In this section, we will summarize data accumulated regarding how direct changes in BH3-only members contribution to carcinogenesis. The best studies to address their contribution are found in experimental mouse knockout models. These models will be compared with data obtained from human studies to also assess their effect on human cancer.

Bim-deficient lymphoid and myeloid cells facilitate tumorigenesis in mice with a tumor-prone background.[78] In one study, half of the bim-null mice became terminally ill within a year.[78] Loss of a single allele of Bim in the presence of a Myc transgene causes robust tumor formation especially in the form of acute B-cell leukemia.[79] This is thought to be due to the observation that apoptosis triggered by Myc is mediated by its induction of Bim.[79] Thus without Bim, Myc tumors have a great survival advantage. Lymphoma from Bim-deficient mice also show resistance to cytokine deprivation,[78] which may help these cells grow unchecked. In human cancer, microarray data showed Bim is down regulated by exhibiting homozygous deletion in mantle cell lymphoma and promoter hypermethylation in Burkitt's lymphoma.[80] Thus Bim plays a significant role in tumor-suppression in controlled mouse models and may indeed be relevant in human cancer.

Aging Bid-deficient mice develop a myeloproliferative disorder and progress to a fatal malignancy resembling chronic myelomonocytic leukemia.[81] In 2 years, 53% of Bid-deficient mice have developed leukemia.[81] Bid-deficient myeloid precursors reportedly show enhanced colony-formation potential in vitro.[81] Thus Bid acts as a tumor suppressor in myeloid cells. The tumorigenetic contribution of Bid in humans is not as evident. An immunohistochemistry survey found increased Bid expression in prostate cancer, colon cancer, ovarian cancer, brain cancer and B-cell non-Hodgkin's lymphoma.[82] Therefore, even though mouse data supports the role of Bid as a tumor-suppressor, its role in human cancer is less evident.

Bad-deficient mice were viable and with age develop diffuse large B-cell lymphoma of germinal center origin.[59] Tumors begin appearing about 15 months of age.[59] Tumor development due to loss of BH3-only proteins generally requires substantial time which suggests that accumulation of other mutations are necessary to induce tumorigenesis. Indeed, increasing mutation rate by sublethally γ-irradiating mice lacking Bad produced greater incidence of leukemia.[59] Loss of Bad also offers advantage for cancer cells by allowing them to become independent of survival signals that control the growth of normal cells. Withdrawal of epidermal growth factor (EGF) increased death of wild-type mammary epithelial cells but Bad null cells were more resistant.[59] Studies in human have found the *Bad* gene is methylated in multiple myeloma cells.[83] Thus Bad loss may be tumorigenic in both mice and humans.

It has been shown that the oncogene E1A induces apoptosis and upregulated Bik though a p53-dependent pathway.[68] However, mice deficient in Bik showed no signs of developing cancer. In human studies, Bik has been found to be methylated in multiple myeloma cells.[83] Immunohistochemistry studies show Bik loss of expression in malignant kidney epithelium compared to their adjacent normal tissue.[84] Thus, Bik is associated with cell death during oncogene activation and cancer progression, but by itself loss of Bik may not be greatly tumorigenic.

Despite its role as an important effector of p53 mediated apoptosis, loss of Puma is not tumorigenic on its own since no spontaneous tumors arise in Puma-deficient mice over time. However, its loss does dramatically reduce the time required for lymphoma formation in Myc transgenic mice.[85] These lymphomas usually maintain wild-type p53 that can still function in inducing cell cycle arrest due to γ-irradiation.[85] Puma loss also reduced time of tumor formation of MEFs transformed by E1A and ras-encoding retroviruses in nude mice.[85] Similarly, loss of Puma greatly promotes apoptotic resistance in MEFs transformed by the Myc oncogene.[72] In humans, no relationship was found between Puma expression level with HNSCC and lung cancer.[86] A correlation was found where Puma decreased upon melanoma progression.[87] Since p53 loss is relatively rare in melanoma,[88] Puma loss in these cells may provide selective advantage from its p53-related activity. Thus, Puma loss may not be highly tumorigenic on its own but may provide many transformed cells with a survival advantage.

Mice deficient in Noxa showed no signs of developing cancer. Noxa was also down regulated by mutation and silenced in diffuse large B-cell lymphoma.[80] Mutational analysis done on colon adenocarcinomas, advanced gastric adenocarcinomas, nonsmall-cell lung carcinomas, breast carcinomas, urinary bladder transitional cell carcinomas and hepatocellular carcinomas found no inactivating mutations associated with these cancers.[89] Thus the evidence available does not conclusively show that Noxa loss significantly contributes to tumorigenesis.

The effect of Bmf on tumorigenesis remains to be determined. Microarray data showed levels of Bmf increase in anoikis conditions. Knock-down of Bmf prevented anoikis and promoted anchorage independent growth.[74] Circumstantial evidence exists to suggest its involvement in cancer since its chromosomal location 15q14 harbors a tumor-suppressor found in patients with advanced breast, lung and brain cancer.[90,91] Yet this evidence is not sufficient to make conclusions regarding the impact of Bmf as a tumor-suppressor.

Mice deficient in Hrk alone showed no signs of developing cancer. However, the *Hrk* gene is often inactivated by methylation in gastric and colorectal cancer compared to normal colon or stomach.[92] Though the gene suppression of this BH3-only member is associated with cancer, these associations alone do not necessarily point to a causal relationship.

Throughout this survey of BH3-only proteins and their impact on cancer development, Bim and Bad show strongest evidence of being human tumor suppressors. In mouse models Bid also is a tumor suppressor but relevance to human cancer is not evident. This does not rule out the possibility that other BH3-only proteins may also be tumor suppressors as well. The weakness of certain BH3-only proteins as tumor-suppressors may be due to their redundancy of function during various apoptotic stimuli. For this reason, it is not surprising that upregulation of anti-apoptotic

BCL-2 family proteins that can block a wide range of BH3-only members have a greater oncogenic ability[93-95] than the loss of individual BH3-only members alone.

BH3-Only Proteins and Chemotherapy Response and Resistance

BH3-only proteins may not only have a profound effect on cancer development but also on cancer treatment. Many cancer therapeutics cause cellular stress that lead to apoptotic death. However, in many cases the exact mechanism of how these therapeutics cause death is not very well studied. The following table summaries some of the BH3-only proteins implicated in various cancer therapeutic treatments.

In most of the studies shown in Table 1, the BH3-only proteins listed was either knocked-down or knocked out. If treatment of the cells deficient in the BH3-only protein of interest resulted in survival then that protein is implicated as a BH3-only protein involved in cell killing. However, since BH3-only protein expression can be tissue specific, the set of BH3-only proteins actually involved in a certain drug-induced killing may be different for cancers of varying origin. Even cancer cells lines from the same tissue type can differ dramatically in regards to the BH3-only proteins that they may employ in cellular suicide. Yet despite these issues, some tentative generalizations can be drawn from our current understanding of BH3-only proteins and their involvement in cancer therapeutics.

One noted trend is the prevalence of Bim involvement in a wide variety of cancer therapy responses. Since Bim is an activator and can also inhibit all anti-apoptotic BCL-2 family proteins, loss of its activity may more noticeably affect drug-induced death than the other BH3-only proteins. Likewise, Puma can inhibit a wide range of anti-apoptotic proteins and is also important in a wide range of cancer therapeutic killing.

One final item worth noting is the fact that each cancer treatment often mobilizes many BH3-only proteins to evoke apoptosis. For example, Imatinib was found to cause cell death through activation of several BH3-only proteins including Bim, Bmf and Bad.[111] Imatinib induced Bim through transcriptional activation and inhibition of its proteasomal degradation.[111] Imatinib also upregulate Bad by dephosphorylation and Bmf though transcriptional induction. Knock-down of Bim caused partial resistance to Imatinib.[117] However, fetal liver cells lacking both Bim and Bad are nearly completely resistant to Imatinib.[111] Thus, the effectiveness of chemotherapeutic treatment relies on many BH3-only proteins to ensure cell death.

Table 1. BH3-only proteins implicated in apoptosis due to various treatments

Cancer Treatment	Cellular Damage	BH3-Only Proteins
5-Fluorouracil	DNA synthesis inhibitor	Bim,[96] Puma,[96] Bad[97] and Bid[57,98]
Adriamycin	DNA damaging agent	Bim,[99] Bik,[70] Bid[100] and Puma[23]
Bortezomib	Proteasome inhibitor	Bim,[101] Noxa[101] and Bik[101]
Etoposide	DNA damaging agent	Bim,[99] Bid,[102,103] Puma[73] and Noxa[73]
Gefitnib	EGFR inhibitor	Bim[104] and Bad[105]
Glucocorticoid	Antagonize immune cells	Bim[106-108] and Puma[107]
HDAC inhibitor	Inhibit histone deacytylse	Bim[109] Bmf[110]
Imatinib	Kinase inhibitor	Bim,[111] Bmf[111] and Bad[111]
Irradiation	DNA damaging agent	Bim,[107] Bid[112] and Puma[73,107]
MEK inhibitors	MAPK inhibitor	Bim[113] and Bmf[113]
Paclitaxel	Microtubule inhibitor	Bim[114-116]

BH3-Only Mimetic Drugs

Due to the significant effect of BH3-only proteins on drug response, various companies have developed small molecules that mimic the BH3 domain and induce apoptotic death. One such drug, ABT-737 is the most potent and specific inhibitor (K_i < 1 nM) of BCL-2, BCL-XL and BCL-2.[118] Studies have shown significant cytotoxic effects of ABT-737 as a single agent on primary follicular lymphoma and B-CLL cells.[118,119] As a single agent ABT-737 is not as effective in solid tumor cell lines but does show activity in small cell lung cancer (SCLC) cells.[118] For a review see Labi et al.[120] In vivo, ABT-737 is not only well tolerated but showed a sixfold increase in event-free survival of immune compromised mice harboring transplanted human pediatric ALL cells when combined with vincristine, dexamethasone and L-ASP.[121]

Based on the activator/sensitizer model, the therapeutic window that allows normal cells to resist ABT-737 treatment and cancer cells to succumb to apoptosis is a phenomenon often termed "BCL-2 addiction." Cancer cells often exhibit characteristics that would evoke the activity of BH3-only proteins like genomic instability and oncogene activation. One way for these cancer cells to survive is to upregulate levels of anti-apoptotic proteins to protect them from death. Indeed it has been shown that many cancer cells over-expressing anti-apoptotic BCL-2 are preloaded with the activator Bim even prior to drug treatment.[119] These cells are in a state that is referred to as "primed" for death and only require sensitizer mediated displacement of activators to trigger cell death.[17] Based on the activator/sensitizer model, ABT-737 would be classified as a sensitizer mimetic since it is not able to activate Bax and Bak on its own. Indeed, ABT-737 has been observed to displace the activator Bim from BCL-2 to cause cytochrome c release.[119] Moreover, the status of being "primed" has been positively correlated with ABT-737 sensitivity.[17,119,122] For this reason, BH3-only mimetics are extremely useful chemotherapeutics against cancer cells.

Despite the effectiveness of ABT-737, its inability to inhibit Mcl-1 can limits its efficacy. Cells displaying resistance to ABT-737 are often shown to be protected by Mcl-1.[123,124] For this reason development of BH3 mimetics which target Mcl-1 and other antiapoptotic proteins are currently explored.

Conclusion

As the sentinels of cellular dysfunction, BH3-only proteins play an important role in the detection and clearance of potentially cancerous cells. A better understanding of BH3-only protein function has already yielded an effective drug ABT-737 that can mimic their activity in the selective killing of cancer cells. However, many solid tumors are not responsive when treated with ABT-737 alone. Since ABT-737 has sensitizer function, it likely requires the presence of other BH3-only proteins to induce cell killing. Indeed, in combination with other chemotherapeutic drugs that can induce activation of other BH3-only proteins, ABT-737 does show greater effectiveness. For this reason, a deeper understanding of BH3-only protein regulation and response to drug action is needed to develop more effective treatment options. Furthermore, understanding how various oncogenes effect the BH3-only proteins mobilization can help physicians tailor drug treatment to the oncogenes involved in individual patient tumors. Standing in the cross roads between detection of cellular derangement and the cellular death machinery, BH3-only proteins serve as important allies in cancer cell killing and rational treatment design.

References

1. Enari M, Sakahira H, Yokoyama H et al. A caspase-activated DNase that degrades DNA during apoptosis and its inhibitor ICAD. Nature 1998; 391(6662):43-50.
2. Khosravi-Far R, Esposti MD. Death receptor signals to mitochondria. Cancer Biol Ther 2004; 3(11):1051-1057.
3. Du C, Fang M, Li Y et al. Smac, a mitochondrial protein that promotes cytochrome c-dependent caspase activation by eliminating IAP inhibition. Cell 2000; 102(1):33-42.
4. Liu X, Kim CN, Yang J et al. Induction of apoptotic program in cell-free extracts: requirement for dATP and cytochrome c. Cell 1996; 86(1):147-157.
5. Wang X. The expanding role of mitochondria in apoptosis. Genes Dev 2001; 15(22):2922-2933.

6. Jiang X, Wang X. Cytochrome c promotes caspase-9 activation by inducing nucleotide binding to Apaf-1. J Biol Chem 2000; 275(40):31199-31203.

7. Rodriguez J, Lazebnik Y. Caspase-9 and APAF-1 form an active holoenzyme. Genes Dev 1999; 13(24):3179-3184.

8. Zou H, Li Y, Liu X et al. An APAF-1.cytochrome c multimeric complex is a functional apoptosome that activates procaspase-9. J Biol Chem 1999; 274(17):11549-11556.

9. Garcia-Saez AJ, Mingarro I, Perez-Paya E et al. Membrane-insertion fragments of Bcl-xL, Bax and Bid. Biochemistry 2004; 43(34):10930-10943.

10. Schlesinger PH, Saito M. The Bax pore in liposomes, Biophysics. Cell Death Differ 2006; 13(8):1403-1408.

11. Wei MC, Zong WX, Cheng EH et al. Proapoptotic BAX and BAK: a requisite gateway to mitochondrial dysfunction and death. Science 2001; 292(5517):727-730.

12. Wei MC, Lindsten T, Mootha VK et al. tBID, a membrane-targeted death ligand, oligomerizes BAK to release cytochrome c. Genes Dev 2000; 14(16):2060-2071.

13. Petros AM, Olejniczak ET, Fesik SW. Structural biology of the BCL-2 family of proteins. Biochim Biophys Acta 2004; 1644(2-3):83-94.

14. Hsu YT, Youle RJ. Bax in murine thymus is a soluble monomeric protein that displays differential detergent-induced conformations. J Biol Chem 1998; 273(17):10777-10783.

15. Petros AM, Nettesheim DG, Wang Y et al. Rationale for Bcl-xL/Bad peptide complex formation from structure, mutagenesis and biophysical studies. Protein Sci 2000; 9(12):2528-2534.

16. Cheng EH, Levine B, Boise LH et al. Bax-independent inhibition of apoptosis by BCL-XL. Nature 1996; 379(6565):554-556.

17. Certo M, Del Gaizo Moore V, Nishino M et al. Mitochondria primed by death signals determine cellular addiction to antiapoptotic BCL-2 family members. Cancer Cell 2006; 9(5):351-365.

18. O'Connor L, Strasser A, O'Reilly LA et al. Bim: a novel member of the BCL-2 family that promotes apoptosis. EMBO J 1998; 17(2):384-395.

19. Wang K, Yin XM, Chao DT et al. BID: a novel BH3 domain-only death agonist. Genes Dev 1996; 10(22):2859-2869.

20. Yang E, Zha J, Jockel J et al. Bad, a heterodimeric partner for BCL-XL and BCL-2, displaces Bax and promotes cell death. Cell 1995; 80(2):285-291.

21. Boyd JM, Gallo GJ, Elangovan B et al. Bik, a novel death-inducing protein shares a distinct sequence motif with BCL-2 family proteins and interacts with viral and cellular survival-promoting proteins. Oncogene 1995; 11(9):1921-1928.

22. Oda E, Ohki R, Murasawa H et al. Noxa, a BH3-only member of the BCL-2 family and candidate mediator of p53-induced apoptosis. Science 2000; 288(5468):1053-1058.

23. Nakano K, Vousden KH. PUMA, a novel proapoptotic gene, is induced by p53. Mol Cell 2001; 7(3):683-694.

24. Inohara N, Ding L, Chen S et al. Harakiri, a novel regulator of cell death, encodes a protein that activates apoptosis and interacts selectively with survival-promoting proteins BCL-2 and Bcl-X(L). EMBO J 1997; 16(7):1686-1694.

25. Puthalakath H, Villunger A, O'Reilly LA et al. Bmf: a proapoptotic BH3-only protein regulated by interaction with the myosin V actin motor complex, activated by anoikis. Science 2001; 293(5536):1829-1832.

26. Zhong Q, Gao W, Du F et al. Mule/ARF-BP1, a BH3-only E3 ubiquitin ligase, catalyzes the polyubiquitination of Mcl-1 and regulates apoptosis. Cell 2005; 121(7):1085-1095.

27. Guo B, Godzik A, Reed JC. Bcl-G, a novel pro-apoptotic member of the BCL-2 family. J Biol Chem 2001; 276(4):2780-2785.

28. Yasuda M, Theodorakis P, Subramanian T et al. Adenovirus E1B-19K/BCL-2 interacting protein BNIP3 contains a BH3 domain and a mitochondrial targeting sequence. J Biol Chem 1998; 273(20):12415-12421.

29. Oberstein A, Jeffrey PD, Shi Y. Crystal structure of the BCL-XL-Beclin 1 peptide complex: Beclin 1 is a novel BH3-only protein. J Biol Chem 2007; 282(17):13123-13132.

30. Liu Z, Lu H, Jiang Z et al. Apolipoprotein l6, a novel proapoptotic BCL-2 homology 3-only protein, induces mitochondria-mediated apoptosis in cancer cells. Mol Cancer Res 2005; 3(1):21-31.

31. Broustas CG, Gokhale PC, Rahman A et al. BRCC2, a novel BH3-like domain-containing protein, induces apoptosis in a caspase-dependent manner. J Biol Chem 2004; 279(25):26780-26788.

32. Mund T, Gewies A, Schoenfeld N et al. Spike, a novel BH3-only protein, regulates apoptosis at the endoplasmic reticulum. FASEB J 2003; 17(6):696-698.

33. Tan KO, Tan KM, Chan SL et al. MAP-1, a novel proapoptotic protein containing a BH3-like motif that associates with Bax through its BCL-2 homology domains. J Biol Chem 2001; 276(4):2802-2807.

34. Letai A, Bassik MC, Walensky LD et al. Distinct BH3 domains either sensitize or activate mitochondrial apoptosis, serving as prototype cancer therapeutics. Cancer Cell 2002; 2(3):183-192.
35. Zong WX, Lindsten T, Ross AJ et al. BH3-only proteins that bind pro-survival BCL-2 family members fail to induce apoptosis in the absence of Bax and Bak. Genes Dev 2001; 15(12):1481-1486.
36. Kuwana T, Mackey MR, Perkins G et al. Bid, Bax and lipids cooperate to form supramolecular openings in the outer mitochondrial membrane. Cell 2002; 111(3):331-342.
37. Lovell JF, Billen LP, Bindner S et al. Membrane binding by tBid initiates an ordered series of events culminating in membrane permeabilization by Bax. Cell 2008; 135(6):1074-1084.
38. Gavathiotis E, Suzuki M, Davis ML et al. BAX activation is initiated at a novel interaction site. Nature 2008; 455(7216):1076-1081.
39. Willis SN, Fletcher JI, Kaufmann T et al. Apoptosis initiated when BH3 ligands engage multiple BCL-2 homologs, not Bax or Bak. Science 2007; 315(5813):856-859.
40. Chipuk JE, Kuwana T, Bouchier-Hayes L et al. Direct activation of Bax by p53 mediates mitochondrial membrane permeabilization and apoptosis. Science 2004; 303(5660):1010-1014.
41. Kim H, Rafiuddin-Shah M, Tu HC et al. Hierarchical regulation of mitochondrion-dependent apoptosis by BCL-2 subfamilies. Nat Cell Biol 2006; 8(12):1348-1358.
42. Pagliari LJ, Kuwana T, Bonzon C et al. The multidomain proapoptotic molecules Bax and Bak are directly activated by heat. Proc Natl Acad Sci USA 2005; 102(50):17975-17980.
43. Cheng EH, Wei MC, Weiler S et al. BCL-2, BCL-X(L) sequester BH3 domain-only molecules preventing BAX- and BAK-mediated mitochondrial apoptosis. Mol Cell 2001; 8(3):705-711.
44. Chen L, Willis SN, Wei A et al. Differential targeting of prosurvival BCL-2 proteins by their BH3-only ligands allows complementary apoptotic function. Mol Cell 2005; 17(3):393-403.
45. O'Reilly LA, Cullen L, Visvader J et al. The proapoptotic BH3-only protein bim is expressed in hematopoietic, epithelial, neuronal and germ cells. Am J Pathol 2000; 157(2):449-461.
46. Marani M, Tenev T, Hancock D et al. Identification of novel isoforms of the BH3 domain protein Bim which directly activate Bax to trigger apoptosis. Mol Cell Biol 2002; 22(11):3577-3589.
47. Adachi M, Zhao X, Imai K. Nomenclature of dynein light chain-linked BH3-only protein Bim isoforms. Cell Death Differ 2005; 12(2):192-193.
48. Puthalakath H, Huang DC, O'Reilly LA et al. The proapoptotic activity of the BCL-2 family member Bim is regulated by interaction with the dynein motor complex. Mol Cell 1999; 3(3):287-296.
49. Dijkers PF, Medema RH, Lammers JW et al. Expression of the pro-apoptotic BCL-2 family member Bim is regulated by the forkhead transcription factor FKHR-L1. Curr Biol 2000; 10(19):1201-1204.
50. Reginato MJ, Mills KR, Paulus JK et al. Integrins and EGFR coordinately regulate the pro-apoptotic protein Bim to prevent anoikis. Nat Cell Biol 2003; 5(8):733-740.
51. Lei K, Davis RJ. JNK phosphorylation of Bim-related members of the Bcl2 family induces Bax-dependent apoptosis. Proc Natl Acad Sci USA 2003; 100(5):2432-2437.
52. Akiyama T, Bouillet P, Miyazaki T et al. Regulation of osteoclast apoptosis by ubiquitylation of proapoptotic BH3-only BCL-2 family member Bim. EMBO J 2003; 22(24):6653-6664.
53. Barry M, Heibein JA, Pinkoski MJ et al. Granzyme B short-circuits the need for caspase 8 activity during granule-mediated cytotoxic T-lymphocyte killing by directly cleaving Bid. Mol Cell Biol 2000; 20(11):3781-3794.
54. Sutton VR, Davis JE, Cancilla M et al. Initiation of apoptosis by granzyme B requires direct cleavage of bid, but not direct granzyme B-mediated caspase activation. J Exp Med 2000; 192(10):1403-1414.
55. Droga-Mazovec G, Bojic L, Petelin A et al. Cysteine cathepsins trigger caspase-dependent cell death through cleavage of bid and antiapoptotic BCL-2 homologues. J Biol Chem 2008; 283(27):19140-19150.
56. Zha J, Weiler S, Oh KJ et al. Posttranslational N-myristoylation of BID as a molecular switch for targeting mitochondria and apoptosis. Science 2000; 290(5497):1761-1765.
57. Sax JK, Fei P, Murphy ME et al. BID regulation by p53 contributes to chemosensitivity. Nat Cell Biol 2002; 4(11):842-849.
58. Kamer I, Sarig R, Zaltsman Y et al. Proapoptotic BID is an ATM effector in the DNA-damage response. Cell 2005; 122(4):593-603.
59. Ranger AM, Zha J, Harada H et al. Bad-deficient mice develop diffuse large B-cell lymphoma. Proc Natl Acad Sci USA 2003; 100(16):9324-9329.
60. Datta SR, Dudek H, Tao X et al. Akt phosphorylation of BAD couples survival signals to the cell-intrinsic death machinery. Cell 1997; 91(2):231-241.
61. del Peso L, Gonzalez-Garcia M, Page C et al. Interleukin-3-induced phosphorylation of BAD through the protein kinase Akt. Science 1997; 278(5338):687-689.
62. Wang HG, Rapp UR, Reed JC. BCL-2 targets the protein kinase Raf-1 to mitochondria. Cell 1996; 87(4):629-638.

63. Harada H, Becknell B, Wilm M et al. Phosphorylation and inactivation of BAD by mitochondria-anchored protein kinase A. Mol Cell 1999; 3(4):413-422.
64. Zha J, Harada H, Yang E et al. Serine phosphorylation of death agonist BAD in response to survival factor results in binding to 14-3-3 not BCL-X(L). Cell 1996; 87(4):619-628.
65. Wang HG, Pathan N, Ethell IM et al. Ca2+-induced apoptosis through calcineurin dephosphorylation of BAD. Science 1999; 284(5412):339-343.
66. Ayllon V, Martinez AC, Garcia A et al. Protein phosphatase 1alpha is a Ras-activated Bad phosphatase that regulates interleukin-2 deprivation-induced apoptosis. EMBO J 2000; 19(10):2237-2246.
67. Chiang CW, Harris G, Ellig C et al. Protein phosphatase 2A activates the proapoptotic function of BAD in interleukin-3-dependent lymphoid cells by a mechanism requiring 14-3-3 dissociation. Blood 2001; 97(5):1289-1297.
68. Mathai JP, Germain M, Marcellus RC et al. Induction and endoplasmic reticulum location of BIK/NBK in response to apoptotic signaling by E1A and p53. Oncogene 2002; 21(16):2534-2544.
69. Verma S, Zhao LJ, Chinnadurai G. Phosphorylation of the pro-apoptotic protein BIK: mapping of phosphorylation sites and effect on apoptosis. J Biol Chem 2001; 276(7):4671-4676.
70. Real PJ, Sanz C, Gutierrez O et al. Transcriptional activation of the proapoptotic bik gene by E2F proteins in cancer cells. FEBS Lett 2006; 580(25):5905-5909.
71. Han J, Flemington C, Houghton AB et al. Expression of bbc3, a pro-apoptotic BH3-only gene, is regulated by diverse cell death and survival signals. Proc Natl Acad Sci USA 2001; 98(20):11318-11323.
72. Jeffers JR, Parganas E, Lee Y et al. Puma is an essential mediator of p53-dependent and -independent apoptotic pathways. Cancer Cell 2003; 4(4):321-328.
73. Villunger A, Michalak EM, Coultas L et al. p53- and drug-induced apoptotic responses mediated by BH3-only proteins puma and noxa. Science 2003; 302(5647):1036-1038.
74. Schmelzle T, Mailleux AA, Overholtzer M et al. Functional role and oncogene-regulated expression of the BH3-only factor Bmf in mammary epithelial anoikis and morphogenesis. Proc Natl Acad Sci USA 2007; 104(10):3787-3792.
75. Imaizumi K, Tsuda M, Imai Y et al. Molecular cloning of a novel polypeptide, DP5, induced during programmed neuronal death. J Biol Chem 1997; 272(30):18842-18848.
76. Sanz C, Benito A, Inohara N et al. Specific and rapid induction of the proapoptotic protein Hrk after growth factor withdrawal in hematopoietic progenitor cells. Blood 2000; 95(9):2742-2747.
77. Hanahan D, Weinberg RA. The hallmarks of cancer. Cell 2000; 100(1):57-70.
78. Bouillet P, Metcalf D, Huang DC et al. Proapoptotic BCL-2 relative Bim required for certain apoptotic responses, leukocyte homeostasis and to preclude autoimmunity. Science 1999; 286(5445):1735-1738.
79. Egle A, Harris AW, Bouillet P et al. Bim is a suppressor of Myc-induced mouse B-cell leukemia. Proc Natl Acad Sci USA 2004; 101(16):6164-6169.
80. Mestre-Escorihuela C, Rubio-Moscardo F, Richter JA et al. Homozygous deletions localize novel tumor suppressor genes in B-cell lymphomas. Blood 2007; 109(1):271-280.
81. Zinkel SS, Ong CC, Ferguson DO et al. Proapoptotic BID is required for myeloid homeostasis and tumor suppression. Genes Dev 2003; 17(2):229-239.
82. Krajewska M, Zapata JM, Meinhold-Heerlein I et al. Expression of BCL-2 family member Bid in normal and malignant tissues. Neoplasia 2002; 4(2):129-140.
83. Pompeia C, Hodge DR, Plass C et al. Microarray analysis of epigenetic silencing of gene expression in the KAS-6/1 multiple myeloma cell line. Cancer Res 2004; 64(10):3465-3473.
84. Sturm I, Stephan C, Gillissen B et al. Loss of the tissue-specific proapoptotic BH3-only protein Nbk/Bik is a unifying feature of renal cell carcinoma. Cell Death Differ 2006; 13(4):619-627.
85. Hemann MT, Zilfou JT, Zhao Z et al. Suppression of tumorigenesis by the p53 target PUMA. Proc Natl Acad Sci USA 2004; 101(25):9333-9338.
86. Hoque MO, Begum S, Sommer M et al. PUMA in head and neck cancer. Cancer Lett 2003; 199(1):75-81.
87. Karst AM, Dai DL, Martinka M et al. PUMA expression is significantly reduced in human cutaneous melanomas. Oncogene 2005; 24(6):1111-1116.
88. Chin L, Merlino G, DePinho RA. Malignant melanoma: modern black plague and genetic black box. Genes Dev 1998; 12(22):3467-3481.
89. Lee SH, Soung YH, Lee JW et al. Mutational analysis of Noxa gene in human cancers. APMIS 2003; 111(6):599-604.
90. Schmutte C, Tombline G, Rhiem K et al. Characterization of the human Rad51 genomic locus and examination of tumors with 15q14-15 loss of heterozygosity (LOH). Cancer Res 1999; 59(18):4564-4569.

91. Wick W, Petersen I, Schmutzler RK et al. Evidence for a novel tumor suppressor gene on chromosome 15 associated with progression to a metastatic stage in breast cancer. Oncogene 1996; 12(5):973-978.

92. Obata T, Toyota M, Satoh A et al. Identification of HRK as a target of epigenetic inactivation in colorectal and gastric cancer. Clin Cancer Res 2003; 9(17):6410-6418.

93. Nieborowska-Skorska M, Hoser G, Kossev P et al. Complementary functions of the antiapoptotic protein A1 and serine/threonine kinase pim-1 in the BCR/ABL-mediated leukemogenesis. Blood 2002; 99(12):4531-4539.

94. Packham G, White EL, Eischen CM et al. Selective regulation of BCL-XL by a Jak kinase-dependent pathway is bypassed in murine hematopoietic malignancies. Genes Dev 1998; 12(16):2475-2487.

95. Zhou P, Levy NB, Xie H et al. MCL1 transgenic mice exhibit a high incidence of B-cell lymphoma manifested as a spectrum of histologic subtypes. Blood 2001; 97(12):3902-3909.

96. Sinicrope FA, Rego RL, Okumura K et al. Prognostic impact of bim, puma and noxa expression in human colon carcinomas. Clin Cancer Res 2008; 14(18):5810-5818.

97. Sinicrope FA, Rego RL, Foster NR et al. Proapoptotic Bad and Bid protein expression predict survival in stages II and III colon cancers. Clin Cancer Res 2008; 14(13):4128-4133.

98. Lee JH, Soung YH, Lee JW et al. Inactivating mutation of the pro-apoptotic gene BID in gastric cancer. J Pathol 2004; 202(4):439-445.

99. Zantl N, Weirich G, Zall H et al. Frequent loss of expression of the pro-apoptotic protein Bim in renal cell carcinoma: evidence for contribution to apoptosis resistance. Oncogene 2007; 26(49):7038-7048.

100. Kohler B, Anguissola S, Concannon CG et al. Bid participates in genotoxic drug-induced apoptosis of HeLa cells and is essential for death receptor ligands' apoptotic and synergistic effects. PLoS ONE 2008; 3(7):e2844.

101. Fennell DA, Chacko A, Mutti L. BCL-2 family regulation by the 20S proteasome inhibitor bortezomib. Oncogene 2008; 27(9):1189-1197.

102. Shelton SN, Shawgo ME, Robertson JD. Cleavage of bid by executioner caspases mediates feed forward amplification of mitochondrial outer membrane permeabilization during genotoxic stress-induced apoptosis in jurkat cells. J Biol Chem 2009.

103. Song G, Chen GG, Chau DK et al. Bid exhibits S phase checkpoint activation and plays a pro-apoptotic role in response to etoposide-induced DNA damage in hepatocellular carcinoma cells. Apoptosis 2008; 13(5):693-701.

104. Cragg MS, Kuroda J, Puthalakath H et al. Gefitinib-induced killing of NSCLC cell lines expressing mutant EGFR requires BIM and can be enhanced by BH3 mimetics. PLoS Med 2007; 4(10):1681-1689; discussion 1690.

105. Gilmore AP, Valentijn AJ, Wang P et al. Activation of BAD by therapeutic inhibition of epidermal growth factor receptor and transactivation by insulin-like growth factor receptor. J Biol Chem 2002; 277(31):27643-27650.

106. Abrams MT, Robertson NM, Yoon K et al. Inhibition of glucocorticoid-induced apoptosis by targeting the major splice variants of BIM mRNA with small interfering RNA and short hairpin RNA. J Biol Chem 2004; 279(53):55809-55817.

107. Erlacher M, Michalak EM, Kelly PN et al. BH3-only proteins Puma and Bim are rate-limiting for gamma-radiation- and glucocorticoid-induced apoptosis of lymphoid cells in vivo. Blood 2005; 106(13):4131-4138.

108. Zhang L, Insel PA. The pro-apoptotic protein Bim is a convergence point for cAMP/protein kinase A- and glucocorticoid-promoted apoptosis of lymphoid cells. J Biol Chem 2004; 279(20):20858-20865.

109. Gillespie S, Borrow J, Zhang XD et al. Bim plays a crucial role in synergistic induction of apoptosis by the histone deacetylase inhibitor SBHA and TRAIL in melanoma cells. Apoptosis 2006; 11(12):2251-2265.

110. Zhang Y, Adachi M, Kawamura R et al. Bmf is a possible mediator in histone deacetylase inhibitors FK228 and CBHA-induced apoptosis. Cell Death Differ 2006; 13(1):129-140.

111. Kuroda J, Puthalakath H, Cragg MS et al. Bim and Bad mediate imatinib-induced killing of Bcr/Abl+ leukemic cells and resistance due to their loss is overcome by a BH3 mimetic. Proc Natl Acad Sci USA 2006; 103(40):14907-14912.

112. Pradhan S, Kim HK, Thrash CJ et al. A critical role for the proapoptotic protein bid in ultraviolet-induced immune suppression and cutaneous apoptosis. J Immunol 2008; 181(5):3077-3088.

113. VanBrocklin MW, Verhaegen M, Soengas MS et al. Mitogen-activated protein kinase inhibition induces translocation of Bmf to promote apoptosis in melanoma. Cancer Res 2009; 69(5):1985-1994.

114. Tan TT, Degenhardt K, Nelson DA et al. Key roles of BIM-driven apoptosis in epithelial tumors and rational chemotherapy. Cancer Cell 2005; 7(3):227-238.

115. Li R, Moudgil T, Ross HJ et al. Apoptosis of nonsmall-cell lung cancer cell lines after paclitaxel treatment involves the BH3-only proapoptotic protein Bim. Cell Death Differ 2005; 12(3):292-303.
116. Janssen K, Pohlmann S, Janicke RU et al. Apaf-1 and caspase-9 deficiency prevents apoptosis in a Bax-controlled pathway and promotes clonogenic survival during paclitaxel treatment. Blood 2007; 110(10):3662-3672.
117. Kuribara R, Honda H, Matsui H et al. Roles of Bim in apoptosis of normal and Bcr-Abl-expressing hematopoietic progenitors. Mol Cell Biol 2004; 24(14):6172-6183.
118. Oltersdorf T, Elmore SW, Shoemaker AR et al. An inhibitor of BCL-2 family proteins induces regression of solid tumours. Nature 2005; 435(7042):677-681.
119. Del Gaizo Moore V, Brown JR, Certo M et al. Chronic lymphocytic leukemia requires BCL2 to sequester prodeath BIM, explaining sensitivity to BCL2 antagonist ABT-737. J Clin Invest 2007; 117(1):112-121.
120. Labi V, Grespi F, Baumgartner F et al. Targeting the BCL-2-regulated apoptosis pathway by BH3 mimetics: a breakthrough in anticancer therapy? Cell Death Differ 2008; 15(6):977-987.
121. Kang MH, Kang YH, Szymanska B et al. Activity of vincristine, L-ASP and dexamethasone against acute lymphoblastic leukemia is enhanced by the BH3-mimetic ABT-737 in vitro and in vivo. Blood 2007; 110(6):2057-2066.
122. Deng J, Carlson N, Takeyama K et al. BH3 profiling identifies three distinct classes of apoptotic blocks to predict response to ABT-737 and conventional chemotherapeutic agents. Cancer Cell 2007; 12(2):171-185.
123. van Delft MF, Wei AH, Mason KD et al. The BH3 mimetic ABT-737 targets selective BCL-2 proteins and efficiently induces apoptosis via Bak/Bax if Mcl-1 is neutralized. Cancer Cell 2006; 10(5):389-399.
124. Konopleva M, Contractor R, Tsao T et al. Mechanisms of apoptosis sensitivity and resistance to the BH3 mimetic ABT-737 in acute myeloid leukemia. Cancer Cell 2006; 10(5):375-388.

CHAPTER 4

Endoplasmic Reticulum Stress and BCL-2 Family Members

Ross T. Weston and Hamsa Puthalakath*

Abstract

In the eukaryotic cell, the endoplasmic reticulum (ER) plays an important role as the site of lipid synthesis, protein folding and protein maturation. Stringent regulation of redox and calcium homeostasis is paramount, failure of which leads accumulation of unfolded and aggregating proteins resulting in a condition known as ER stress. Eukaryotic cells deal with ER stress by eliciting the unfolded protein response (UPR). This pathway splits into two streams depending on the severity and longevity of the ER stress, where the cell must make a choice for the good of the organism between survival and programmed cell death. The BCL-2 family of proteins is central to the cell death arm of the UPR pathway. This chapter discusses the recent findings on the involvement of BCL-2 family members in the apoptotic process initiated by ER stress and a related process called autophagy. Understanding the molecular mechanisms involved in ER stress and autophagy could have a profound implications developing new therapies for many ER stress associated diseases and cancer.

Introduction

The acquisition of the membrane bound organelles during the evolution of eukaryotes represents one of the fundamental shifts in biochemical reactions; from the relics of prokaryotes in which biochemical processes occur in the cytosol requiring the primordial, anaerobic reducing conditions to the far more sophisticated metabolic pathways in which oxygen is an absolute necessity. The Endoplasmic Reticulum (ER) and mitochondrion are two such organelles eukaryotes acquired early during evolution. The ER is thought to have originated by a fusion between an archaebacterium and a gram-negative eubacterium.[1]

Mitochondrial evolution has no such archaebacterial ancestry, rather has a mono phylogenetic origin from within eubacteria.[2] Through reductive evolution, the relationship between mitochondria and the host cell became interdependent symbiosis.[3] In eukaryotes, the mitochondria provide the energy that the cell needs through oxidative phosphorylation whereas the endoplasmic reticulum plays an entirely different role. It is recognized as the site of synthesis and folding of secreted, membrane-bound and some organelle-targeted proteins. Despite these divergent origin and functions, these two organelles have inter-connected, overlapping functions in regulating cellular calcium homeostasis. This is achieved through the action of a family of protein called Mitofusins, which helps these two organelles to juxtapose and interconnect to regulate intracellular calcium (Ca^{2+}) homeostasis.[4] The overlapping function between these two organelles also extends to an important biological process called apoptosis or programmed cell death (PCD). Release of calcium

*Corresponding Author: Hamsa Puthalakath—Department of Biochemistry, School of Molecular Sciences, La Trobe University, Bundoora, Australia 3086. Email: h.puthalakath@latrobe.edu.au

BCL-2 Protein Family: Essential Regulators of Cell Death, edited by Claudio Hetz.
©2010 Landes Bioscience and Springer Science+Business Media.

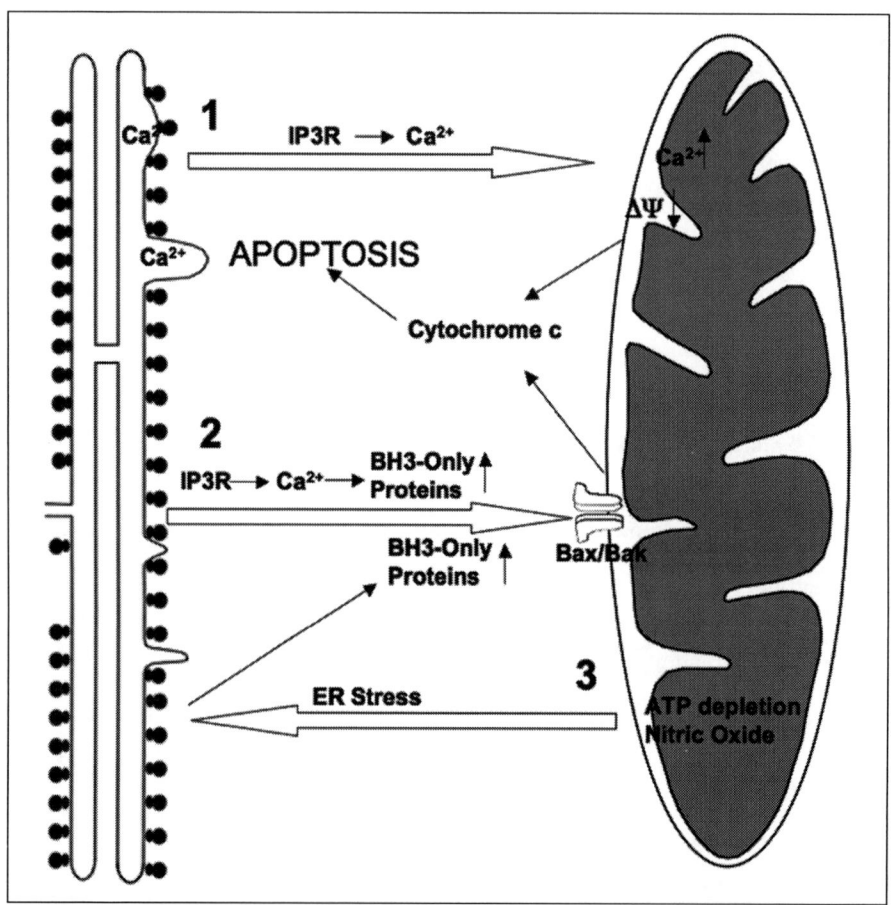

Figure 1. Cross-talk between the ER and the mitochondrion. 1) IP3R opening results in Ca²⁺ release from the ER lumen resulting mitochondrial calcium spike. This could result in the loss of mitochondrial membrane potential, cytochrome release and apoptosis. 2) Loss of Ca²⁺ from the ER lumen triggers ER stress and induction of BH3-only proteins, which results in apoptosis. 3) Loss of ATP generation from the mitochondrion results in misfolding of the proteins in the ER, which leads to apoptosis.

from the ER mediated by activation of inositol 1,4,5-triphosphate (IP3) or ryanodine receptor (RyR), collectively termed as (IP3R) results in spike in cytosolic Ca²⁺. This in turn results in an increase in the mitochondrial Ca²⁺ pool. A precipitous accumulation of Ca²⁺ in mitochondria may lead to the opening of the "transition pore" and accelerate ion exchanges across the inner mitochondrial membrane. As a result, there is a loss of negative membrane potential ($\Delta\psi$) that contributes to the release of apoptotic factors, such as cytochrome c, leading to apoptosis (Fig 1).[5] Apart from this direct effect on mitochondrial apoptosis through perturbed calcium homeostasis, in many physiological contexts, the decision by a cell to live or die is made at the mitochondrial surface based on the stoichiometry of pro- and anti-apoptotic BCL-2 family members. In this scenario as well, calcium flux can play an important role. For example, high affinity engagement of MHC-bound self antigen to the T-cell receptor results in transient calcium release from ER lumen and eventually capacitive calcium flux, which leads to the induction of the pro-apoptotic BCL-2 family member Bim,[6] a process by which autoreactive thymocytes are eliminated through apoptosis. Conversely, optimal mitochondrial function is necessary for optimal functioning of ER. Several factors are required for optimal protein folding in the ER, including ATP, Ca^{2+} and an oxidizing environment to allow disulphide-bond formation.[7] As a consequence of these special requirements, the ER is highly sensitive to stresses that perturb cellular energy levels, the redox state or Ca^{2+} concentration. This is best illustrated by the finding that endogenously generated nitric oxide (NO), which is a pleiotropic signalling molecule, can irreversibly bind to cytochrome c, thus disrupting respiratory chain which eventually results in ER stress.[8] Thus, apart from the physical interconnection between these two divergent organelles,[9] there is a substantial functional overlap and a homeostatic interdependence between these two, which can prove decisive in determining the fate of the cell. The main focus of this review is how events at the ER i.e., ER stress or the Unfolded Protein Response (UPR) play a role in mitochondrial apoptotic pathway through the regulation of BCL-2 family members.

UPR: The Good and the Evil Side

The endoplasmic reticulum (ER) is the hub of protein trafficking inside the cell. Secreted, membrane bound and organelle targeted proteins are all channelled through the ER before being glycosylated and folded to make them competent to be exported. The concentration of proteins in the lumen of ER can reach the threshold of "solvent crisis" (i.e., 100 mg/ml, see ref. 10) where keeping proteins in a folded soluble conformation is achieved at a premium cost i.e., the requirement of a considerable amount of cellular energy. For example, the most abundant ER chaperone, BiP/GRP78, uses the energy from ATP hydrolysis to promote folding and prevent aggregation of proteins within the ER. Apart from the energy requirement, chaperones can have other requirements; BiP/GRP78 for example needs optimal glucose concentration whereas chaperones such as calnexin and calreticulin require optimal concentration of Ca^{2+} for their function.[11] Furthermore, optimal functioning of protein disulfide isomerases is necessary to catalyze and monitor disulfide bond formation in a regulated and ordered manner. As a consequence of these special requirements, the ER is highly sensitive to stresses that perturb cellular energy levels, the redox state, Ca^{2+} concentration and glucose homeostasis. Such stresses reduce the protein folding capacity of the ER, which can result in the accumulation and aggregation of unfolded proteins and/or an imbalance between the load of resident and transit proteins in the ER and the organelle's ability to process that load. This condition is referred to as ER stress. However, in most cases, cell ensures that this stress does not overwhelm the protein-folding capacity by activating the pathway that is referred to as the unfolded protein response (UPR). UPR can promote cellular repair and sustained survival by reducing the load of unfolded proteins through global attenuation of protein synthesis and/or up-regulation of chaperones, enzymes and structural components of the ER.[10] This response is mediated through three ER trans-membrane receptors: pancreatic ER kinase (PKR)-like RE kinase (PERK), activating transcription factor 6 (ATF6) and inositol requiring enzyme 1 (IRE1). In resting cells, all of these three ER stress receptors are maintained in an inactive state through their association with the ER chaperone, GRP78. Upon accumulation of unfolded proteins, GRP78 dissociates from these three receptors, which leads to their activation and initiation of the UPR. Thus, the UPR is a pro-survival response to reduce the accumulation of unfolded proteins and restore normal ER function.[12] However, when misfolded protein aggregation persists and the ER stress cannot be resolved, signalling switches from a pro-survival to a pro-apoptotic response. This arm of the ER stress response is mainly mediated through the transcription factor CHOP.[13] This bifurcation of signal transduction is manifested in the divergent role of ER stress/UPR plays in various patho-physiological and developmental contexts.

The diseases associated with malfunctioning of the ER collectively known as Endoplasmic Reticulum Storage Diseases (ERSDs) can arise for a variety of reasons and can occur in any cell type, thus explaining the wide variety of the clinical spectrum of ER stress associated diseases. Impaired lysosomal protein degradation due to defective or missing hydrolases results in an expansion of lysosomes filled with indigestible degradation intermediates that can inhibit normal cellular function causing diseases such as Fabry's, Tay-Sachs, Gaucher's and Niemann-Pick diseases.[14] However, defective lysosomal degradation can also lead to protein accumulation in the ER and ER stress because proteins targeted to lysosomes are channelled through ER after Mannose-6-phosphate modification in the Golgi. Misfolded protein accumulation in the ER can also occur due to genetic mutation. For example, mutations in the signal sequence affecting membrane translocation or signal processing are suspected to cause familial hypoparathyroidism, coagulation factor X deficiency, Crigler-Najjar syndrome Type II, familial central diabetes insipidus or chronic pancreatitis (please see ref. 14 and references there in). Mutations in the coding region can make proteins inherently misfolded. This could be manifested as autosomal dominant or recessive trait depending on whether they are functional as monomers or multimers. These include Cystic Fibrosis or CF,[15] α1-antitrypsin deficiency,[16] thyroglobulin deficiency,[17] Diabetes insipidus[18] to name a few. In such cases, disease may result from the mere lack of the mutant protein in question and/or may be caused indirectly by toxic effects of the misfolded protein or aggregates thereof on the cell and subsequent cell death. The direct connection between ER stress and apoptosis has not been studied in these above-mentioned diseases. However, there are mouse models where the link between ER stress and apoptosis has been established as the aetiology of the disease. One of the earliest such studies was by Miller's group.[19] Expression of class I major histocompatibility complex in islet beta cells of transgenic mice resulted in islet destruction and diabetes. Electron micrograph studies of such islets showed that in these cells there was clear evidence of defective endoplasmic reticulum. In case of the *"akita"* mouse, a spontaneous mutation in the insulin-2 gene (Ins2) (Cys96Tyr) is responsible for the diabetic phenotype.[20] This mutation disrupts a disulphide bond formation between the A chain (A7) and the B chain (B7) of proinsulin, resulting in misfolded insulin protein being retained in the ER leading to beta cell apoptosis. In a more recent study, Hetz et al, using a transgenic mouse model, expressing a mutant version of superoxide dismutase-1 (SOD1), which is associated with familial amyotrophic lateral sclerosis (FALS) demonstrated the link between ER stress and apoptosis in neuronal cells.[21] The direct link between ER stress and apoptosis has been reported in a limited number of human diseases such as hereditary hemochromatosis, sporadic idiopathic pulmonary fibrosis and in human temporal lobe epilepsy.[22-24]

Apart from its association with various pathophysiologies, the UPR plays a critical role in certain developmental processes that are associated with increased demand for protein synthesis and/or export, such as differentiation of immunoglobulin (Ig)-secreting plasma cells and CD[8]-T-lymphocytes which develop into cytokine and cytolytic factor secreting killer cells.[25,26] The UPR also plays an important role in myoblast formation as reported by Nakanishi et al.[27] Increased ER stress enhanced differentiation-associated apoptosis of myoblasts. It is believed that apoptosis induced by ER stress selectively eliminates vulnerable cells leaving myoblast that are more resistant to apoptosis which, eventually fuse to form myofibres. Furthermore, the UPR is also necessary for the development and differentiation of the professional antigen presenting cells such as dendritic cells.[28] In reconstitution experiments, chimeric mice lacking the transcription factor XBP-1 in the lymphoid compartment had decreased numbers of both conventional and plasmacytoid DCs with reduced survival both at baseline level and in response to TLR signalling. *Thus, the paradox of life as we know it, life and death, spirit and matter, real and unreal, UPR seems to have both a good and an evil side!*

UPR and BCL-2 Family Members

When misfolded protein accumulation is overwhelming, cells switch from survival mode to apoptotic mode. Although the mechanism of this mode of apoptosis was of great debate and conjecture for some time, a clear picture is emerging. Some of the earliest works linking an ER stress molecule to apoptosis came from David Ron's Lab.[29] However, the literature started getting inundated with papers describing the role of caspase-12 in this process. Jin Yin Yuan's group from Harvard Medical School first described the role of caspase-12 in ER-specific apoptosis pathway and its contribution to amyloid-beta neurotoxicity.[30] In this paper, it was shown that caspase-12 is localized to the ER and activated by ER stress, including disruption of ER calcium homeostasis and accumulation of excess proteins in ER, but not by membrane- or mitochondrial-targeted apoptotic signals. Using gene knockout mice, they had reported that mice that were deficient in caspase-12 were resistant to ER stress-induced apoptosis, but their cells underwent apoptosis in response to other death stimuli. The activated caspase-12 was reported to be able to cleave pro-caspase-9, which in turn activated caspase-3 and this resulted in apoptosis.[31] However, the lack of active caspase-12 in the majority of the human population posed a problem for this model but Hitomi et al[32] argued that the human equivalent of mouse caspase-12, caspase-4 was processed during ER stress and was able to induce apoptosis. This intriguing result had a surprising caveat i.e., the apoptotic pathway initiated by caspase-4 could be blocked by ectopic expression of BCL-2 suggesting the involvement of a BH3-only protein or activation of pro-apoptotic proteins bax and bak by caspase-4 activation.

The notion of direct activation of caspases during ER stress by caspase-12 was laid to rest through the elegant experiments from Nicholson's lab.[33] Caspase-12 is a polymorphic allele in humans with majority of the population carry a mutation resulting in a truncated protein. However, in a sub population of African descent, the read-through single nucleotide polymorphism results in the production of a full-length procaspase called Csp12-L, rendering those individuals susceptible to sepsis.[34] This observation was followed by gene knock-out studies which unequivocally demonstrated that indeed, caspase-12 is involved in dampening the production of pro-inflammatory cytokines such as interleukin (IL)-1β, IL-18 (interferon (IFN)-γ inducing factor) and IFN-γ. Caspase-12 knockout mice were resistant to septic shock and peritonitis. Most importantly, ER stress-induced apoptosis was normal in these mice.[33] This work was followed by a recent publication that showed that recombinant caspase-12 could mediate auto proteolytic maturation of its own proenzyme, in both *cis* and *trans*, it was not able to cleave any other polypeptide substrate, including other caspase proenzymes (including caspase-9), apoptotic substrates, cytokine precursors, or proteins in the endoplasmic reticulum that normally undergo caspase-mediated proteolysis.[35]

The earlier reports on regulation of ER stress induced apoptosis by BCL-2 family members derived from the observation that in many cells, BCL-2 is localized to ER membranes in addition to the mitochondria and nuclear envelop.[36] The notion that ER targeted BCL-2 somehow regulates ER stress induced apoptosis was bolstered by the fact that ectopic expression of BCL-2 with ER targeting sequence of cytochrome B5 (Bcl-Cb5) protected against many apoptotic stimuli except etoposide and serum withdrawal (please see ref. 37 and references therein). However, with the benefit of hindsight and in light of recent findings, it is reasonable to suggest that BCL-2 located at the ER membrane acts as a sink for the pro-apoptotic molecules that are activated/induced during ER stress including BH-3 only proteins and Bax/Bak group of proteins. The differential protection by Bcl-Cb5 could be explained by the fact that different BH3-only proteins and Bax/Bak have different affinities for the anti-apoptotic members of the BCL-2 family and these proteins are differentially induced/activated in response to different stimuli.[38,39]

One of the first convincing experiments that linked ER stress with BCL-2 family came from both Guido Kroemer's group[40] and Craig Thompson's group.[41] It was demonstrated that in ER stress mediated apoptosis, elicited by tunicamycin, thapsigargin or brefeldin A, mitochondrial membrane potential (MMP) played a critical role and this could be altered by the ectopic expression of BCL-XL. Similarly, using *bax−/−/bak−/−* fibroblasts, Zong et al[41] had shown that these two proteins are critical for ER stress induced apoptosis as well as apoptosis induced by various other agents. Though the conventional wisdom based on structural studies suggests that these two proteins exert their effect through their association with the mitochondria, recent studies suggest that they associate with the ER membrane and influence ER stress induced apoptosis. In different cell types including HeLa, MCF7, 293T and MEFs, ER stress resulted in the translocation of Bak and Bax to ER membranes and this lead to the loss of luminal Ca^{2+}.[42] Furthermore and *bax−/−, bak−/−* mouse embryonic fibroblasts have a reduced resting concentration of calcium in the ER.[43] This is attributed to the modulation of inositide 3-phosphate receptor (IP3R) function by BCL-2, which in turn is regulated by Bax and Bak subfamily of pro-apoptotic BCL-2 family members.[44] Similarly, Hetz et al[45] have shown that Bax and Bak can regulate ER stress response through modulating the function of the ER membrane kinase/nuclease, IRE1α. IRE1α is the nuclease that is involved in the alternate splicing of XBP-1 protein, which is one of the important modulators of UPR. By generating a conditional double deletion of *bax* and *bak* in the liver, these authors have reported that DKO mice responded abnormally to tunicamycin-induced ER stress in the liver, with extensive tissue damage and decreased expression of the IRE1 substrate X-box—binding protein 1 and its target genes. However, apoptosis was significantly reduced in the liver suggesting an additional role for the Bax/Bak subfamily in modulating ER stress response. This is consistent with our own finding that tunicamycin induced apoptosis was significantly reduced in *bim*[-/-] mice without any liver damage, perhaps due to the presence of Bax and Bak in these mice.[46]

The BH3-only members of the BCL-2 family act as sentinels that selectively trigger apoptosis in response to developmental cues or stress-signals such as ER stress. Widely expressed mammalian BH3-only proteins are thought to act by binding to and neutralizing their pro-survival counterparts such as BCL-2, BCL-2XL and Mcl1. Activation of BH3-only proteins directly or indirectly results in the activation of proapoptotic BAX and BAK to trigger cell death. PUMA was one of the first BH3-only proteins implicated in ER stress induced apoptosis.[47] In human SH-SY5Y neuroblastoma cells, treatment with tunicamycin resulted in the transcriptional induction of PUMA. This was necessary and sufficient to induce apoptosis. However, this induction was only partially p53 dependant because PUMA induction was only delayed, not completely abrogated in SAOS-2 (p53 deficient) cells. The role of PUMA in ER stress induced apoptosis seems to be restricted to cell type and tissue type. While most of the initial studies were done in cells of neuronal origin, where other BH3-only proteins such as Bim did not seem to have any role[47] subsequent studies have shown that PUMA plays role in ER stress induced apoptosis in other tissues such as cardiomyocytes,[48] melanocytes[49] and in MEFs.[50] PUMA also has been shown to play a role during the early stages if chronic neurodegenerative diseases such as amyotrophic lateral sclerosis (ALS). Using the SOD1 (G93A) mouse model as well as human post mortem samples from ALS patients, Kieran et al[51] have reported that increased ER stress and defects in protein degradation in motoneurons during disease progression. Genetic deletion of puma significantly improved motoneuron survival and delayed disease onset and motor dysfunction in SOD1 (G93A) mice. However, it had no significant effect on lifespan, suggesting that other ER stress-related cell-death proteins may play a role. Indeed, Hetz et al,[21] using the same mouse model, have shown that up regulation of Bim results in motor neuron loss in ALS mouse models and the disease onset was significantly delayed in *bim*[-/-] mice. Results from our lab show that in many tissue types, including breast epithelial cells, thymocytes, macrophages and kidney epithelial cells, ER stress induces Bim.[46] In thymocytes and epithelial cells, ER stress induced apoptosis is Bim dependant whereas PUMA did not seem to play any significant role. This induction had two components. Transcriptional induction mediated by ER stress induced transcription factor CHOP and posttranslational stabilization of Bim by protein phosphatase 2A (PP2A). Bim is also induced in PC12 cells, which are of rat adrenal medullar origin (sympathetic nervous system), in response to ER stress.[52] Thus both Bim and PUMA seem to have overlapping roles in ER stress induced apoptosis in neuronal cells whereas in tissues such as thymocytes, only Bim appears to be important and in fibroblasts (MEFs), PUMA is the major contributor and Bim having little role to play. Apart from Bim and PUMA, other BH3-only proteins are reported to be involved in ER stress induced apoptotic pathway. Noxa is induced in p53 dependant manner in MEFs[50] and in neuroblastoma and melanoma cells in response to fenretinide and emphasise.[53] Bid is cleaved by caspase 2 in response to heat shock, which is known to induce the unfolded protein response and bid null MEFs were resistant to heat shock induced apoptosis.[54,#265] Bid is also reported to be activated in response to ER stress induced by homoharringtonine in MUTZ-1 leukemic cells.[55] However, Bid deficiency did not have any effect on ER stress induced apoptosis in HeLa cells.[56]

Apart from the direct transcriptional activation and or posttranslational modification of BCL-2 family members, BCL-2 family members can be regulated through additional intermediate signalling pathways in response to ER stress/UPR signalling. JNK activation is often associated with ER stress[57] and this could lead to either transcriptional activation or posttranslational modification of BCL-2 family members. JNK is reported to transcription-ally activate Bak and Bim[58] and also has been reported to mediate translocation of Bax from the cytosol to the mitochondria by phosphorylating 14-3-3.[59] Further, JNK activation could result in the translocation of Bim and Bmf from their respective cytoskeletal complexes to mitochondria.[60] Similarly, Glycogen synthase kinase 3β (GSK3β) is activated in response to ER stress by dephosphorylation at Ser[9] by PP2A.[61,62] GSK3β could potentially phosphorylate MCL-1 and Bax leading to degradation of the former and activation of the latter.[63,64] However, the caveat to the above suggestions is that activation of JNK and GK3β during ER stress is well documented but its role in regulating BCL-2 family members during ER stress is only a possibility that no one has proven so far!

In summary, these findings highlight the fact that ER stress could initiate different signal-ling pathways depending on the type of cells/tissue and or the agent that is used to induce the stress and duration of the stress which, in turn will determine which BCL-2 family member is regulated. For example, macrophages from *bim*[−/−] mice are protected against thapsigargin-induced apoptosis whereas Bim deficiency does not offer protection against tunicamycin induced apoptosis in this cell type (H. Puthalakath, unpublished observation).

Autophagy, UPR and BCL-2 Family Members

Autophagy is a process by which intracellular organelles are targeted for lysosomal degrada-tion, which provides a source of amino acids under starvation conditions.[65] Similar to UPR, autophagy initiation could have dichotomous outcomes as different as starvation adaptation and apoptosis (Fig 2). In between these two extremities, it also has an important role in vari-ous pathological and patho-physiological contexts such as intracellular protein and organelle clearance, development, anti-aging, elimination of micro-organisms, tumour suppression and antigen presentation.[66]

In yeast, about 31 *atg* gene products regulate the process of autophagy. *ATG6* or Beclin-1 (*BECNI*) was the first mammalian autophagy gene identified as a haplo-insufficient tumor sup-pressor.[67] Beclin-1 was initially identified as a BCL-2 interacting protein in a yeast two-hybrid screen.[68] The interaction between the anti apoptotic protein, BCL-2 and the autophagy protein, Beclin 1, represents an important point of convergence of the autophagic machinery, apoptosis and the BCL-2 family of proteins. However, little was known about the functional significance of this interaction for considerable time. Eventually, Beth Levin's group demonstrated that BCL-2 plays an important role in Beclin-1 mediated autophagy.[69] Wild-type BCL-2 protein, but not Beclin 1 binding defective mutants of BCL-2, inhibited Beclin 1-dependent autophagy in yeast and mammalian cells. Similarly, cardiac BCL-2 transgenic expression inhibited autophagy in mouse heart muscle. Furthermore, Beclin 1 mutants that could not bind to BCL-2 induced more autophagy than wild-type Beclin 1 and, unlike wild-type Beclin 1, promoted cell death. Thus, BCL-2 not only functions as an anti apoptotic protein, but also as an anti-autophagy protein via its inhibitory interaction with Beclin 1. This anti-autophagy function of BCL-2 may help maintain autophagy at levels that are compatible with cell survival, rather than cell death.[69]

Apart from the interaction between Beclin 1 and BCL-2, there are many other instances where a direct relationship between autophagy and apoptosis has been demonstrated. Inhibition of autophagy enhances drug-induced apoptosis in Myc-lymphoma cells and in hepatocarcinoma cells.[70,71]

Conversely, inhibition of apoptosis enhances autophagy and autophagy induced cell death. *bax/bak*[−/−] cells or cells over expressing BCL-X$_L$ were defective in undergoing apoptosis but were more radiosensitive than the WT cells in autophagy.[72,73] Thus, these results provide a demonstrable link between autophagy and apoptosis. Recent results also show that in many instances, this link is established through induction of the UPR. In the yeast, S. cerevisiae, Hac1 (the homologue of mammalian Xbp1) can activate several of autophagy genes including Atg5, 7, 8 and 19.[74] Since many proteins that are targeted to lysosomes are mannose 6-phosphate glycosylated in the Golgi compartment, it is conceivable that any defect in lysosomal processing or inhibition

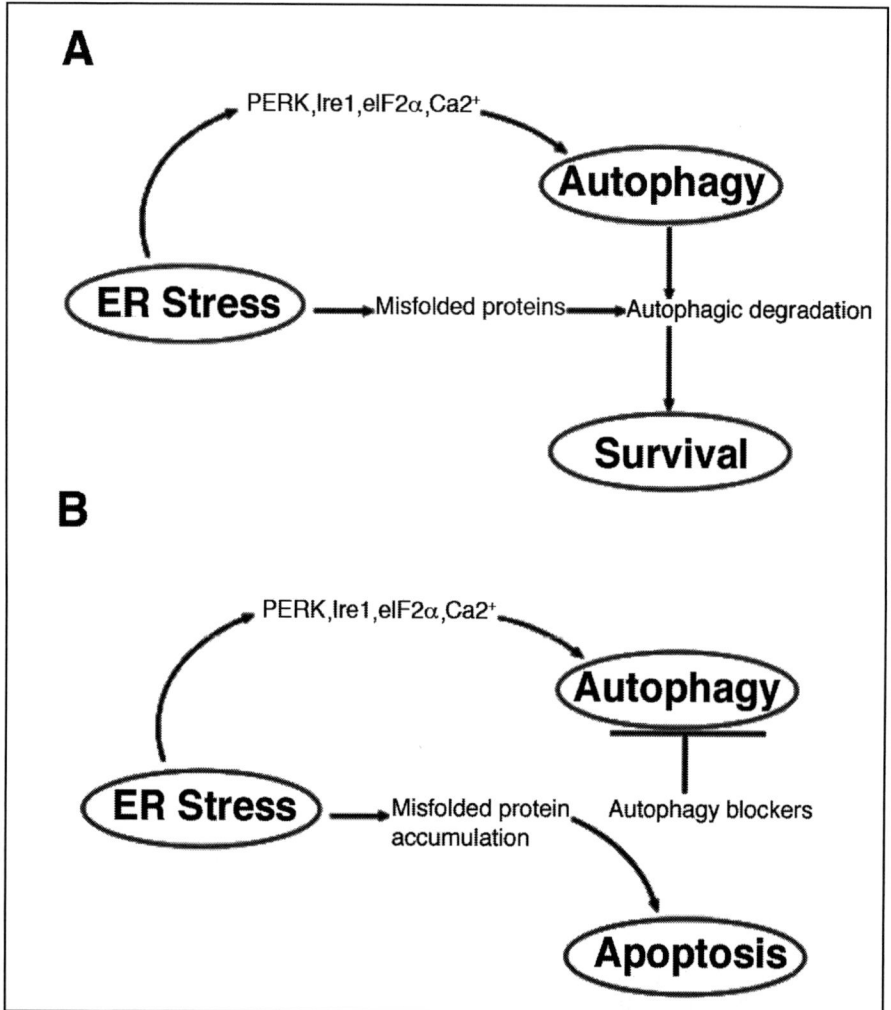

Figure 2. Mutual regulation of the ER stress pathway and the autophagic pathway. A) ER stress induces autophagy pathway through PERK, IRE1, eIF2α and increased Ca²⁺, which helps to degrade unfolded protein and alleviate ER stress leading to survival. B) Blockage of autophagy results in the exacerbation of ER stress resulting apoptosis.

of autophagy could result in the accumulation of proteins in the secretory pathway and thereby, induce UPR/ER stress.[75] Huntington's disease results in the cytoplasmic accumulation of Polyglutamine repeats (PolyQ) and can lead to a global attenuation of proteasomal activity. This inturn leads to the accumulation of misfolded proteins in the ER and UPR and trigger autophagy.[76] PERK, IRE1, eIF2α and increased Ca²⁺ have been implicated as mediators of ER stress-induced autophagy in mammalian cells.[74,77-80] Similar to Hac1 in yeast, PERK, IRE1, eIF2α and Ca²⁺ all induce various autophagy genes in mammalian cells. However, Ca²⁺ flux can also induce DAP kinase and calpain protease activities that are known to be involved in autophagy regulation.[81,82] ER localised BCL-2 can modulate the Ca²⁺ flux and thereby regulate the autophagy response independent of its binding to Beclin 1.[78] These results suggest that induction of the UPR could trigger an autophagy response, which may help cells to counterbalance the endoplasmic reticulum expansion during the unfolded protein response and maintain a new steady-state level of ER abundance even in the face of continuously accumulating unfolded proteins.[83] On the other hand, blocking autophagy can enhance ER stress- induced cell death,[74,77,83] although this does not happen in all cases and although it seems counterintuitive, autophagy itself in certain situations can also trigger ER stress which can inturn lead to apoptosis. This was illustrated by Altman et al[84] who reported that autophagy was induced in hematopoietic cells by growth factor withdrawal. This lead to apoptosis mediated through CHOP-dependant induction of the BH3-only protein Bim. Thus, UPR and autophagy mechanisms appear to be intricately linked with interdependent controls where altered activity of one system impinges upon the other. The regulatory network appears to be controlled by bi-directional hierarchical and modular mechanisms, the precise nature of which has still to be worked out.

Conclusion

The endoplasmic reticulum represents the cell's quality control site for accurate folding of secretory and membrane proteins. Accurate folding is achieved through the co-ordinated function of a large number of proteins including chaperones and protein peptide isomerases and various glycosyltransferases. It is also the site of cells drug metabolism mediated through the Cytochrome P450 family of proteins[85] and a major pool of cellular calcium. Several factors are required the optimal functioning of ER including ATP, Ca^{2+} and an oxidising environment. As a consequence of these special requirements, the ER is highly sensitive to stresses that perturb cellular energy levels, the redox state or Ca^{2+} concentration. Malfunctioning of the ER is implicated in the aetiology of many diseases. Such disease can be environmentally driven such as high fat diet in obesity and diabetes[85] or be the result of genetic mutations, in which mutant proteins fail to pass the ER quality control.[14] Therefore, understanding of the UPR/ ER stress signal transduction mechanisms is of paramount importance in treating these diseases as illustrated by Ozcan et al.[86] In the last 10 years, it has also become clear that the UPR plays an important role in the survival of malignant cancers and therefore targeting the UPR with chemotherapeutic drugs offers a great potential in treating this disease.[87]

The signal transduction pathway involved in autophagy is less well understood compared to UPR. Recently, it has become clear that akin to UPR, tumor cells use autophagy as a mean to survive under conditions of metabolic stress, hypoxia and perhaps even chemotherapy. Although these two systems may function independently from each other, there are also important connections with interdependent controls, where altered activity of one system impinges upon the other. Treatment of tumor cells with drugs that are able to specifically trigger further ER-stress (or block autophagy) could result in two outcomes; (i) such drugs by themselves might result in increased ER-stress resulting in antitumor activity, or (ii) the ER-stress thus generated could bring down the cells' survival threshold and make them more sensitive towards conventional chemotherapeutic agents. One main common denominator to both UPR and autophagy induced apoptosis is the role of BCL-2 family proteins. It is now being recognized as an important player in the life-and-death decisions of the ER-stress response and autophagy. Recent literature has provided many examples of ER stress/autophagy inducing agents being used in chemotherapy where apoptosis is initiated through BH3-only BCL-2 family members. These include Velcade,[88,89] Nonsteroidal anti-inflammatory drugs,[90] HDAC inhibitors such as Tubacin[91] and inhibitors of lysosomal degradation such as Chloroquine.[92] This approach will be particularly useful in treating multiple myeloma and endocrine tumors where cells secrete large quantities of proteins which should make it easier to induce ER stress and in treating solid tumors where hypoxic conditions prime the cells for ER stress induced apoptosis and in treating p53 deficient tumors.

Acknowledgement

HP is supported by the National Health and Medical Research Council (Australia).

References

1. Gupta RS. Life's Third Domain (Archaea): An established fact or an endangered paradigm? A new proposal for classification of organisms based on protein sequences and cell structure. Theoretical Population Biology 1998; 54(2):91-104.
2. Gray MW BG, Lang BF. Mitochondrial evolution. Science 1999; 283(5407):1476-1481.
3. Andersson SG KC. Reductive evolution of resident genomes. Trends Microbiol 1998; 6(7):263-268.
4. de Brito OM SL. Mitofusin 2 tethers endoplasmic reticulum to mitochondria. Nature 2008; 456(7222):605-610.
5. Kanwar YS SL. Shuttling of calcium between endoplasmic reticulum and mitochondria in the renal vasculature. Am J Physiol Renal Physiol 2008; 295(5):F101-F102.
6. Canté-Barrett K, Gallo EM, Winslow MM et al. Thymocyte negative selection is mediated by protein kinase C- and Ca^{2+}-dependent transcriptional induction of bim. J Immunol 2006; 176(4):2299-2306.
7. Gaut JR HL. The modification and assembly of proteins in the endoplasmic reticulum. Curr Opin Cell Biol 1993; 5(4):589-595.
8. Xu W LL, Charles IG, Moncada S. Nitric oxide induces coupling of mitochondrial signalling with the endoplasmic reticulum stress response. Nat Cell Biol 2004; 6(11):1129-1134.
9. Csordás G RC, Várnai P, Walter L et al. Structural and functional features and significance of the physical linkage between ER and mitochondria. J Cell Biol 2006; 174(7):915-921.
10. Kaufmann RJ. Orchestrating the unfolded protein response in health and disease. J Clin Invest 2002; 110:1389-1398.
11. Molinari M HA. Chaperone selection during glycoprotein translocation into the endoplasmic reticulum. Science 2000; 288(5464):331-333.
12. Schröder M KR. The mammalian unfolded protein response. Annu Rev Biochem 2005; 74:739-789.
13. Di Sano F FE, Tufi R, Achsel T et al. Endoplasmic reticulum stress induces apoptosis by an apoptosome-dependent but caspase 12-independent mechanism. J Biol Chem 2006; 281(5):2693-2700.
14. Rutishauser J SM. Endoplasmic reticulum storage diseases. Swiss Med Wkly 2002; 132(17-18):211-222.
15. Cheng SH GR, Marshall J, Paul S et al. Defective intracellular transport and processing of CFTR is the molecular basis of most cystic fibrosis. Cell 1990; 63(4):827-834.
16. Le A SJ, Ferrell GA, Shaker JC et al. Association between calnexin and a secretion-incompetent variant of human alpha 1-antitrypsin. J Biol Chem 1994; 269(10):7514-7519.
17. Kim PS KO, Arvan P. An endoplasmic reticulum storage disease causing congenital goiter with hypothyroidism. J Cell Biol 1996; 133(3):517-527.
18. Mulders SM BD, Rijss JP, Kamsteeg EJ et al. An aquaporin-2 water channel mutant which causes autosomal dominant nephrogenic diabetes insipidus is retained in the Golgi complex. J Clin Invest 1998; 102(1):57-66.
19. Allison J ML, Culvenor J, Bartholomeusz RK et al. Overexpression of beta 2-microglobulin in transgenic mouse islet beta cells results in defective insulin secretion. Proc Natl Acad Sci USA 1991; 88(6):2070-2074.
20. Yoshioka M, Kayo T, Ikeda T et al. A novel locus, Mody4, distal to D7Mit189 on chromosome 7 determines early-onset NIDDM in nonobese C57BL/6 (Akita) mutant mice. Diabetes 1997; 46(5):887-894.
21. Hetz C, Thielen P, Fisher J et al. The proapoptotic BCL-2 family member BIM mediates motoneuron loss in a model of amyotrophic lateral sclerosis. Cell Death Differ 2007; 14(7):1386-1389.
22. de Almeida SF, M. dS. The unfolded protein response in hereditary haemochromatosis. J Cell Mol Med 2007; 12(2):421-434.
23. Korfei M RC, Mahavadi P, Henneke I et al. Epithelial endoplasmic reticulum stress and apoptosis in sporadic idiopathic pulmonary fibrosis. Am J Respir Crit Care Med 2008; 178(8):838-846.
24. Yamamoto A MN, Schindler CK, So NK et al. Endoplasmic reticulum stress and apoptosis signaling in human temporal lobe epilepsy. J Neuropathol Exp Neurol 2006; 65(3):217-225.
25. Brunsing R OS, Weber F, Bicknell A et al. B- and T-cell development both involve activity of the unfolded protein response pathway. J Biol Chem 2008; 283(26):17954-17961.
26. Huse M QE, Davis MM. Shouts, whispers and the kiss of death: directional secretion in T-cells. Nat Immunol 2008; 9(10):1105-1111.
27. Nakanishi K DN, Morishima N. Endoplasmic reticulum stress increases myofiber formation in vitro. FASEB J 2007; 21(11):2994-3003.

28. Iwakoshi NN PM, Glimcher LH. The transcription factor XBP-1 is essential for the development and survival of dendritic cells. J Exp Med 2007; 204(10):2267-2275.
29. Zinszner H KM, Wang X, Batchvarova N et al. CHOP is implicated in programmed cell death in response to impaired function of the endoplasmic reticulum. Genes Dev 1998; 12(7):982-995.
30. Nakagawa T ZH, Morishima N, Li E et al. Caspase-12 mediates endoplasmic-reticulum-specific apoptosis and cytotoxicity by amyloid-beta. Nature 2000; 403(6765):98-103.
31. Morishima N NK, Takenouchi H, Shibata T et al. An endoplasmic reticulum stress-specific caspase cascade in apoptosis. Cytochrome c-independent activation of caspase-9 by caspase-12. J Biol Chem 2002; 277(37):34287-34294.
32. Hitomi J KT, Eguchi Y, Kudo T et al. Involvement of caspase-4 in endoplasmic reticulum stress-induced apoptosis and Abeta-induced cell death. J Cell Biol 2004; 165(3):347-356.
33. Saleh M MJ, Wolinski MK, Bensinger SJ et al. Enhanced bacterial clearance and sepsis resistance in caspase-12-deficient mice. Nature 2006; 440(7078):1064-1068.
34. Saleh M VJ, Graham RK, Huyck M et al. Differential modulation of endotoxin responsiveness by human caspase-12 polymorphisms. Nature 2004; 429(6987):75-79.
35. Roy S SJ, Houde C, Loisel TP et al. Confinement of caspase-12 proteolytic activity to autoprocessing. Proc Natl Acad Sci USA 2008; 105(11):4133-4138.
36. Krajewski S TS, Takayama S, Schibler MJ et al. Investigation of the subcellular distribution of the bcl-2 oncoprotein: residence in the nuclear envelope, endoplasmic reticulum and outer mitochondrial membranes. Cancer Res 1993; 53(19):4701-4714.
37. Thomenius MJ DC. BCL-2 on the endoplasmic reticulum: protecting the mitochondria from a distance. J Cell Sci 2003; 116:4493-4499.
38. Chen L WS, Wei A, Smith BJ et al. Differential targeting of prosurvival BCL-2 proteins by their BH3-only ligands allows complementary apoptotic function. Mol Cell 2005; 17(3):393-403.
39. Willis SN CL, Dewson G, Wei A et al. Proapoptotic Bak is sequestered by Mcl-1 and Bcl-xL, but not BCL-2, until displaced by BH3-only proteins. Genes Dev 2005; 19(11):1294-1305.
40. Boya P, Cohen I, Zamzami N et al. Endoplasmic reticulum stress-induced cell death requires mitochondrial membrane permeabilization. Cell Death Differ 2002; 9(4):465-467.
41. Zong WX LT, Ross AJ, MacGregor GR et al. BH3-only proteins that bind pro-survival BCL-2 family members fail to induce apoptosis in the absence of Bax and Bak. Genes Dev 2001; 15(12):1481-1486.
42. Zong WX LC, Hatzivassiliou G, Lindsten T et al. Bax and Bak can localize to the endoplasmic reticulum to initiate apoptosis. J Cell Biol 2003; 162(1):59-69.
43. Scorrano L OS, Opferman JT, Cheng EH et al. BAX and BAK regulation of endoplasmic reticulum Ca^{2+}: a control point for apoptosis. Science 2003; 300(5616):135-139.
44. Oakes SA SL, Opferman JT, Bassik MC et al. Proapoptotic BAX and BAK regulate the type 1 inositol trisphosphate receptor and calcium leak from the endoplasmic reticulum. Proc Natl Acad Sci USA 2004; 102(1):105-110.
45. Hetz C BP, Fisher J, Lee AH et al. Proapoptotic BAX and BAK modulate the unfolded protein response by a direct interaction with IRE1alpha. Science 2006; 312(5773):572-576.
46. Puthalakath H ORL, Gunn P, Lee L et al. ER stress triggers apoptosis by activating BH3-only protein Bim. Cell 2007; 129(7):1337-1349.
47. Reimertz C, Kogel D, Rami A et al. Gene expression during ER stress-induced apoptosis in neurons: induction of the BH3-only protein Bbc3/PUMA and activation of the mitochondrial apoptosis pathway. J Cell Biol 2003; 162(4):587-97.
48. Nickson P TA, Erhardt P. PUMA is critical for neonatal cardiomyocyte apoptosis induced by endoplasmic reticulum stress. Cardiovasc Res 2007; 73(1):48-56.
49. Jiang CC LK, Avery-Kiejda KA, Wade M et al. Up-regulation of Mcl-1 is critical for survival of human melanoma cells upon endoplasmic reticulum stress. Cancer Res 2008; 68(16):6708-6717.
50. Li J LB, Lee AS. Endoplasmic reticulum stress-induced apoptosis: multiple pathways and activation of p53-up-regulated modulator of apoptosis (PUMA) and NOXA by p53. J Biol Chem 2006; 281(11):7260-7270.
51. Kieran D, Woods I, Villunger A et al. Deletion of the BH3-only protein puma protects motoneurons from ER stress-induced apoptosis and delays motoneuron loss in ALS mice. Proc Natl Acad Sci USA 2007; 104(51):20606-20611.
52. Szegezdi E HK, Kavanagh ET, Samali A et al. Nerve growth factor blocks thapsigargin-induced apoptosis at the level of the mitochondrion via regulation of Bim. J Cell Mol Med 2008. (Epub ahead of print).
53. Armstrong JL VG, Redfern CP, Lovat PE. Role of Noxa in p53-independent fenretinide-induced apoptosis of neuroectodermal tumours. Apoptosis 2007; 12(3):613-622.

54. Bonzon C B-HL, Pagliari LJ, Green DR et al. Caspase-2-induced apoptosis requires bid cleavage: a physiological role for bid in heat shock-induced death. Mol Biol Cell 2006; 17(5):2150-2157.
55. Jie H DH, Xingkui X, Liang G et al. Homoharringtonine-induced apoptosis of MDS cell line MUTZ-1 cells is mediated by the endoplasmic reticulum stress pathway. Leuk Lymphoma 2007; 48(5):964-977.
56. Köhler B AS, Concannon CG, Rehm M et al. Bid participates in genotoxic drug-induced apoptosis of HeLa cells and is essential for death receptor ligands' apoptotic and synergistic effects. PLoS ONE 2008; 3(7):e2844.
57. Ozcan U CQ, Yilmaz E, Lee AH et al. Endoplasmic reticulum stress links obesity, insulin action and type 2 diabetes. Science 2004; 306(5696):457-461.
58. Jin HO PI, An S, Lee HC et al. Up-regulation of Bak and Bim via JNK downstream pathway in the response to nitric oxide in human glioblastoma cells. J Cell Physiol 2006; 206(2):477-486.
59. Tsuruta F, JS, Mori Y et al. JNK promotes Bax translocation to mitochondria through phosphorylation of 14-3-3 proteins. EMBO J 2004; 23:1889-1899.
60. Lei K DR. JNK phosphorylation of Bim-related members of the Bcl2 family induces Bax-dependent apoptosis. Proc Natl Acad Sci USA 2003; 100(5):2432-2437.
61. Resende R FE, Pereira C, Oliveira CR. ER stress is involved in Abeta-induced GSK-3beta activation and tau phosphorylation. J Neurosci Res 2008; 86(9):2091-2099.
62. Song L DSP, Jope RS. Central role of glycogen synthase kinase-3beta in endoplasmic reticulum stress-induced caspase-3 activation. J Biol Chem 2002; 277(47):44701-44708.
63. Opferman J. Unraveling MCL-1 degradation. Cell Death Differ 2006; 13(8):1260-1262.
64. Linseman DA BB, Precht TA, Phelps RA et al. Glycogen synthase kinase-3beta phosphorylates Bax and promotes its mitochondrial localization during neuronal apoptosis. J Neurosci 2004; 24(44):9993-10002.
65. Mizushima N. Autophagy: process and function. Genes Dev 2007; 21(22):2861-2873.
66. Mizushima N. The pleiotropic role of autophagy: from protein metabolism to bactericide. Cell Death Differ 2005; 2:1535-1541.
67. Yue Z JS, Yang C, Levine AJ et al. Beclin 1, an autophagy gene essential for early embryonic development, is a haploinsufficient tumor suppressor. Proc Natl Acad Sci USA 2003; 100(25):15077-15082.
68. Liang XH KL, Jiang HH, Gordon G et al. Protection against fatal Sindbis virus encephalitis by beclin, a novel BCL-2-interacting protein. J Virol 1998; 72(11):8586-8596.
69. Pattingre S TA, Qu X, Garuti R et al. BCL-2 antiapoptotic proteins inhibit Beclin 1-dependent autophagy. Cell 2005; 122(6):927-939.
70. Amaravadi RK YD, Lum JJ, Bui T et al. Autophagy inhibition enhances therapy-induced apoptosis in a Myc-induced model of lymphoma. J Clin Invest 2007; 117(2):326-336.
71. Longo L PF, Scardino A, Alabiso O et al. Autophagy inhibition enhances anthocyanin-induced apoptosis in hepatocellular carcinoma. Mol Cancer Ther 2008; 7(8):2476-2485.
72. Kim KW MR, Cao C, Albert JM et al. Autophagy for cancer therapy through inhibition of pro-apoptotic proteins and mammalian target of rapamycin signaling. J Biol Chem 2006; 281(48):36883-36890.
73. Shimizu S KT, Mizushima N, Mizuta T et al. Role of BCL-2 family proteins in a non-apoptotic programmed cell death dependent on autophagy genes. Nat Cell Biol 2004; 6(12):1221-1228.
74. Yorimitsu T NU, Yang Z, Klionsky DJ. Endoplasmic reticulum stress triggers autophagy. J Biol Chem 2006.
75. Dahms NM LP, Kornfeld S. Mannose 6-phosphate receptors and lysosomal enzyme targeting. J Biol Chem 1989; 262(21):12115-12118.
76. Momoi T. Conformational diseases and ER stress-mediated cell death: apoptotic cell death and autophagic cell death. Curr Mol Med 2006; 6(1):111-118.
77. Ogata M HS, Saito A, Morikawa K et al. Autophagy is activated for cell survival after endoplasmic reticulum stress. Mol Cell Biol 2006; 26(24):9220-9231.
78. Høyer-Hansen M BL, Szyniarowski P, Campanella M et al. Control of macroautophagy by calcium, calmodulin-dependent kinase kinase-beta and BCL-2. Mol Cell 2007; 25(2):193-205.
79. Kouroku Y FE, Tanida I, Ueno T et al. ER stress (PERK/eIF2alpha phosphorylation) mediates the polyglutamine-induced LC3 conversion, an essential step for autophagy formation. Cell Death Differ 2007; 14(2):230-239.
80. Fujita E KY, Isoai A, Kumagai H et al. Two endoplasmic reticulum-associated degradation (ERAD) systems for the novel variant of the mutant dysferlin: ubiquitin/proteasome ERAD(I) and autophagy/lysosome ERAD(II). Hum Mol Genet 2007; 16(6):618-629.
81. Demarchi F BC, Copetti T, Tanida I et al. Calpain is required for macroautophagy in mammalian cells. J Cell Biol 2006; 175(4):595-605.

82. Inbal B BS, Sabanay I, Shani G et al. DAP kinase and DRP-1 mediate membrane blebbing and the formation of autophagic vesicles during programmed cell death. J Cell Biol 2002; 157(3):455-468.
83. Bernales S MK, Walter P. Autophagy counterbalances endoplasmic reticulum expansion during the unfolded protein response. PLoS Biol 2006; 4(12):e423.
84. Altman BJ WJ, Zhao Y, Coloff JL et al. Autophagy provides nutrients but can lead to chop-dependent induction of bim to sensitize growth factor deprived cells to apoptosis. Mol Biol Cell 2008. (Epub ahead of print).
85. Coon MJ. Cytochrome P450: nature's most versatile biological catalyst. Annu Rev Pharmacol Toxicol 2005; 45:1-25.
86. Ozcan U YE, Ozcan L, Furuhashi M et al. Chemical chaperones reduce ER stress and restore glucose homeostasis in a mouse model of type 2 diabetes. Science 2006; 313(579):1137-1140.
87. Strasser A Puthalakath H. Fold up or perish: unfolded protein response and chemotherapy. Cell Death Differ 2008; 15(2):223-225.
88. Paoluzzi L GM, Bhagat G, Furman RR et al. The BH3-only mimetic ABT-737 synergizes the antineoplastic activity of proteasome inhibitors in lymphoid malignancies. Blood 2008; 12(7):2906-2916.
89. J. Adams MK. Development of the proteasome inhibitor Velcade (Bortezomib). Cancer Invest 2004; 22:304-311.
90. Tsutsumi S GT, Tomisato W, Mima S et al. Endoplasmic reticulum stress response is involved in nonsteroidal anti-inflammatory drug-induced apoptosis. Cell Death Differ 2004; 11(9):1009-1016.
91. Hideshima T BJ, Wong J, Chauhan D et al. Small-molecule inhibition of proteasome and aggresome function induces synergistic antitumor activity in multiple myeloma. Proc Natl Acad Sci USA 2005; 102(24):8567-8572.
92. Maclean KH, Dorsey FC, JL L et al. Targeting lysosomal degradation induces p53-dependent cell deah and prevents cancer in mouse models of lymphomagenesis. J Clin Invest 2008; 118(1):79-88.

CHAPTER 5

Targeting Survival Pathways in Lymphoma

Luca Paoluzzi and Owen A. O'Connor*

Abstract

Targeting cellular death pathways including apoptosis is a promising strategy for cancer drug discovery. To date at least three major types of cell death have been distinguished, including: apoptosis, autophagy, and necrosis. Increasing evidence has begun to support a role of Bcl-2–family members in the cellular pathways involved in each of these processes. The induction of apoptosis in different types of tissue and in response to various stressors is a complex process that is controlled by different BCL-2 family members. Pharmacologic modulation of BCL-2 proteins and apoptosis can be achieved through different ways including the use of: (1) Modified peptides; (2) Small molecule inhibitors of anti-apoptotic proteins; (3) Antisense strategies; and (4) TRAIL targeting. Non-peptide based small-molecule inhibitors of signaling pathways are at present the strategy of choice given their low antigenicity and generally more favorable pharmacokinetic and pharmacodynamic features, especially as they pertain to volume of distribution and intracellular accumulation. Bcl2-family inhibitors are showing impressive preclinical efficacy in animal models and are moving rapidly towards phase I and II clinical trials. Appropriate preclinical studies will need to identify the optimal strategies for combining these agents, with an emphasis on the importance of dose and schedule dependency.

Introduction

The pathways responsible for adult tissue homeostasis are governed significantly though not exclusively by BCL-2—family proteins. To date at least three major types of cell death have been distinguished, including: apoptosis, autophagy and necrosis. Increasing evidence has begun to support a role of BCL-2—family members in the cellular pathways involved in each of these processes.[1]

In 1988 Vaux et al discovered that BCL-2 provides a distinct survival signal to the cell and may contribute to transformation by allowing a clone to persist until other oncogenes, like c-myc for example, become activated.[2] BCL-2 cooperates with c-myc to promote proliferation of B-cell precursors, some of which can become tumorigenic. At that time apoptosis had already been recognized as an intrinsic cellular program that plays a complementary role to mitosis in regulating tissue homeostasis.[3] It wasn't until the 1990's however that important studies in the biology of apoptosis revealed: (1) the marked evolutionary conservation of the apoptotic machinery; (2) the central role of a class of cysteine proteases, later called caspases; (3) the existence of pro- and anti-apoptotic relatives of BCL-2 characterized by complex member

*Corresponding Author: Owen A. O'Connor—Department of Medicine and Pharmacology and Department of Clinical Research and Cancer Treatment, New York Univeristy Cancer Institute, Division of Hematological Malignancies and Medical Oncology, The New York University Langone Medical Center, New York, New York. Email: owen.o'connor@nyumc.org

BCL-2 Protein Family: Essential Regulators of Cell Death, edited by Claudio Hetz.
©2010 Landes Bioscience and Springer Science+Business Media.

interactions and (4) the existence of at least two distinct apoptotic pathways in mammalian cells, one involving the mitochondria ("intrinsic" pathway) and the other involving the death receptors ("extrinsic" pathway).[4-7]

Mitochondria play a central role in apoptosis by housing and releasing the proteins that participate in caspase activation as well as the inhibitors of caspases and are well acknowledged mediators of necrotic cell death.[8] For example, defects in electron chain transport in respiring mitochondria cause lipid peroxidation and consequent membrane damage from reactive oxygen species, which impairs normal ion-homeostasis, causing cellular swelling, plasma and lysosomal membrane rupture leading to release of hydrolytic enzymes that degrade intracellular proteins, nucleic acids and lipids. BCL-2—family proteins also modulate the pathways leading to necrotic cell death, largely orchestrated also by the mitochondria.[9] The point of regulation may be linked to the ability of BCL-2—family proteins to control the outer mitochondrial membrane permeability: loss of cytochrome c from mitochondria in fact, can interrupt the electron chain transport and is heralded as one of the key steps leading to the initiation of programmed cell death.[10] The release of several other proteins that contribute to 'non-apoptotic' cell death, including DNAse, endonuclease G and apoptosis-inducing factor, a flavoprotein reported to enter the nucleus and promote genomic degradation, are additional consequences of the mitochondrial outer membrane permeabilization (MOMP).

BCL-2—family proteins have recently shown to be involved in a third process leading to death called autophagy, an evolutionarily conserved response for catabolizing macromolecules and organelles during prolonged periods of nutrient deprivation. Autophagy (literally translated to mean 'self-eating') can generate the substrates required to maintain ATP synthesis. Autophagy is initially induced to prolong cell survival, but when taken to extremes, it causes cell death. BCL-2 and BCL-X$_L$ suppress autophagy by binding the protein Beclin (ATG7), an essential component of the mammalian autophagy system that marks autophagic vesicles for fusion with lysosomes for digestion and recycling of intracellular components.[11] The anti-autophagic functions of BCL-2 are not centralized to the mitochondria, but rather, appear to be regulated from the endoplasmic reticulum (ER), where a considerable proportion of anti-apoptotic BCL-2 and related proteins often resides.[12] In this regard BCL-2—family proteins have regulatory effects on several proteins and processes in the ER, including those influencing the unfolded protein response (UPR). The UPR is an evolutionarily conserved adaptive response that detects accumulation of unfolded proteins in the lumen of the ER, causing induction of chaperones and transporters that are intended to restore homeostasis through the retrograde transport of unfolded proteins into the cytosol for eventual ubiquitination and protolytic degradation.[13] Chaperone-mediated autophagy complements proteasome-dependent degradation of misfolded proteins and represents another major route of disposal of defective proteins. Persistent ER stress induces both caspase-dependent and caspase-independent cell death programs, which are modulated by BCL-2—family proteins.

BCL-2 Proteins and Apoptosis

All BCL-2 family members contain characteristic regions of homology termed BH (BCL-2 Homology) domains.[14] Members of this family can be divided into three groups based on their structures and functions. The anti-apoptotic (pro-survival) group, including BCL-2, BCL-X$_L$, Mcl-1, BCL-2 and A1, contain four BH domains and antagonize the pro-apoptotic function of BAK and BAX.[15] The second group, including BAX and BAK, are pro-apoptotic and contain multiple BH domains.[14] Cells in which both the pro-apoptotic BAX and BAK have been deleted are essentially resistant to apoptosis.[16] The third group is also considered pro-apoptotic, though are called the "BH3-only proteins". In mammals, this group includes at least eight members (BAD, BID, BIK, BIM, BMF, HRK, NOXA and PUMA) that display sequence homology with other BCL-2 family members only within the amphipathic and α-helical BH3 segments.[17] The multiple BH3-only proteins are thought to fine tune the apoptotic response in mammalian cells. Structural studies have revealed that the BH1, BH2 and BH3 domains in the anti-apoptotic proteins fold into a globular domain containing a hydrophobic groove on its surface.[18] The α-helical BH3 domains of pro-apoptotic proteins bind to this hydrophobic groove, in effect 'neutralizing' the anti-apoptotic proteins.[19] In healthy cells, basal levels of anti-apoptotic proteins prevent BAX and BAK from being activated. Upon initiation of an apoptotic signal, BH3-only proteins are activated and competitively bind to the hydrophobic grooves of the anti-apoptotic proteins through the BH3 domains.[20] This serves to displace BAX and BAK and allows them to form multimers which lead to increased permeabilization of the mitochondrial outer membrane.[21] Most if not all apoptotic signals transmitted by BH3 domains converge through BAX and BAK.[22] Once a cell becomes committed to apoptosis, a cascade of irreversible downstream events are triggered to execute cell death, including collapse of the mitochondrial membrane potential, release of the apoptogenic mitochondrial proteins such as cytochrome c, SMAC/Diablo and AIF and activation of caspases.[23,24] The relative levels of BH3-only proteins and their pro-survival relatives is crucial in establishing the threshold for commitment of a cell to apoptosis and, therefore, for the control of tissue homeostasis. The three dimensional structure of BCL-X_L alone and in complex with a peptide from the BAK BH3 domain have laid the foundation for many subsequent mechanistic studies, leading to our present understanding of the models used to appreciate the complexities of this life-death switch.[25]

The BH3 proteins have different capabilities of inducing apoptosis depending on the cell type, owing to their differential binding specificities for various anti-apoptotic proteins.[26-28] The BH3 domains of PUMA and BIM, for example, can bind to all five anti-apoptotic proteins. In contrast, those of BAD and BMF preferentially interact with BCL-2, BCL-X_L, BCL-2, but not with Mcl-1 or A1. The BH3 domains of BID, BIK and HRK strongly bind to BCL-X_L, BCL-2 and A1, but only weakly to BCL-2 and Mcl-1. The BH3 domain of NOXA binds to Mcl-1 and A1, but not to other anti-apoptotic proteins.[26,27] On the other hand, some evidence suggests that several BH3-only proteins such as BID and BIM directly bind to and activate BAX and BAK.[27,29] An alternative hierarchical model among BH3-only proteins has been proposed, in which BID, BIM and perhaps PUMA, are involved in directly activating BAX and BAK, while other BH3-only proteins function by engaging pro-survival proteins, thereby liberating these proteins to interact with BAX and BAK.[30,31] However, emerging evidence indicates that all BH3-only proteins indirectly activate BAX and BAK by binding to anti-apoptotic proteins, as apoptosis induced by many stimuli is intact in mice deficient in both BID and BIM.[32] Additionally, different stress signals activate different BH3-only proteins. For example, NOXA[33] and PUMA[34] are triggered by the p53 pathway. By contrast, BIM, in addition to its major role in haematopoietic homeostasis,[35] is the primary mediators of endoplasmic reticulum stress.[36]

BCL-2 proteins are also involved in the "extrinsic" pathway of apoptosis that is centered on the role of TRAIL (Tumor necrosis factor-related apoptosis-inducing ligand) and its receptors. TRAIL induces apoptosis by binding to the DR4 and DR5 receptors, which causes the intracellular death domains of these receptors to trimerize. This aggregation of membrane proteins leads to the recruitment of FADD and activation of caspase 8, caspase 3 and caspase 7. Caspase 8 activation further amplifies the death signal by activating the intrinsic apoptosis pathway through cleaving the BCL2 family member BID. Cleaved BID binds to BAX and BAK, causing the release of cytochrome c and SMAC from the mitochrondria. This results in the activation of caspase 9 and, subsequently, other downstream caspases.[37]

Pharmacologic Modulation of BCL-2 Proteins and Apoptosis

The induction of apoptosis in different types of tissue and in response to various stressors is a complex process that is controlled by different BCL-2 family members. For example, knockout studies have demonstrated that BCL-X_L is essential in embryogenesis for survival of erythroid progenitors and neuronal cells[38] and that BCL-2 is required for survival of mature T- and B-lymphocytes.[39] BCL-X_L has been found to be essential for regulating platelet survival.[40] Mcl-1 has been demonstrated to be required for implantation and survival of hematopoietic stem cells and progenitor B- and T-lymphoid cells,[41] while BCL-2 sustains developing sperm cells.[42] So, the consequences of blocking the function of individual pro-survival proteins are likely to be very cell-type and target-specific, as antagonists of these family members might well be expected to cause a variety of side effects from lymphopenia to azospermia and thrombocytopenia. Clearly, understanding the underlying biology in both normal and malignant cells will help us understand the side effects of BCL-2 family member antagonists.

Modified Peptides

One innovative strategy to modulate the apoptotic threshold is to design small peptides to fit into the appropriate grooves of the various BCL-2 family members. To date, several short peptides representing the BH3 domain of BH3-only proteins have been designed. The sometimes poor binding affinity of these peptides is likely related in part to their lack of secondary structure (i.e., degree of helicity), though new efforts oriented toward understanding the impact of secondary structure are now being integrated into more novel platforms in peptide design and tailoring.

One interesting approach has been published by Walensky and colleagues. Walensky et al recently used a chemical strategy, termed hydrocarbon stapling, to generate BH3 peptides with improved and well prescribed pharmacologic properties.[43] The stapled peptides, called "stabilized alpha-helix of BCL-2 domains" (SAHBs), retain helical secondary structure, are protease-resistant and cell-permeable molecules that are capable of binding with increased affinity to multidomain BCL-2 member pockets. They recently demonstrated that a SAHB of the BH3 domain from the BID protein specifically activates apoptosis in myeloid leukemia cells and is capable of inhibiting the growth of human leukemia in in vivo xenograft models.

Small Molecule Inhibitors of Anti-Apoptotic Proteins

A number of small molecules targeting anti-apoptotic proteins such as BCL-2 and BCL-X_L have been identified through a variety of methods, including computational modeling, structure-based design and high-throughput screening of natural product and synthetic libraries. Nonpeptide based small-molecule inhibitors of signaling pathways are at present the strategy of choice given their low antigenicity and generally more favorable pharmacokinetic and pharmacodynamic features, especially as they pertain to volume of distribution and intracellular accumulation. In addition, the ease and plasticity of molecular modification allows for facile tailoring of putative drug candidates. These latter features can facilitate and hasten the ability to improve route-specific delivery, increase bioavailability and increase target affinity. While this is a relatively new area, there are now a host of small molecules that have been undergoing lead optimization, early preclinical studies and for some, recent early Phase 1 experiences.[44] Some of these small molecules are discussed below (Tables 1 and 2).

HA14-1 and Analogs

HA14-1 is a synthetic chromene molecule capable of disrupting the BAX—BCL-2 interactions, promoting mitochondrial dysfunction and cytochrome c release.[45,46] It was been the first

Table 1. Examples of novel agents targeting survival pathways

Mechanism	Drugs	References
Inhibition of BCL-2, BCL-X_L, Mcl-1	(-)-Gossypol/AT-101	60-65
Inhibition of BCL-2, BCL-X_L, BCL-2	ABT-737, ABT-263	81-102
Inhibition of BCL-2, BCL-X_L, BCL-2, Mcl-1, A1, Bcl-b	GX015-070/obatoclax mesylate	74-80
Antisense oligonucleotide targets BCL-2 mRNA	Oblimersen sodium/Genasense	104-108
DR4 and DR5 activation through antibodies	Mapatumumab, Apomab, CS-1008	109-111
TRAIL receptor activation	rhApo2L/TRAIL	113-119

Table 2. BCL-2 targeted agents in clinical trial

Agent	Development Status	Dose and Schedule	Dose Limiting Toxicities/ Toxicities	Activity (Disease)
ABT-263[102]	Phase 1 and 2 (Hematologic malignancies)	Escalating	DLT: Elevated LFTs, Thrombocytopenia Arrhytmia	SLL/CLL, NK-T
GX015-070/ obatoclax mesylate[78-80]	Phase 1 and 2 (Hematologic malignancies)	Up to 60 mg, 24h infu-sion, Q2wks or 45 mg/d 1,4,8,11	DLT: QT prolongation. Other toxicities:neurological (eupho-ria, gait disturbance, headache, dizziness), GI (nausea, diarrhea), edema, weight loss, chills, hy-perhydrosis, febrile neutropenia, cough, chest pain	AML, MDS
AT-101[65]	Phase 2 (+ Rituximab)	AT-101 inter-mittent (80 mg d1-3, 15-17 Q28d) versus continuous (30 mg/d for 3 weeks)	No DLT Other toxicities: GI (LFTs)	CLL
Oblimersen sodium[107,108]	Phase 2 and 3 (+ Rituximab or fludarabine +cyclophosph-amide)	3 mg/kg/day as a 7-day continuous IV infusion	Thrombocytopenia, anemia, neutropenia, fatigue, oedema, rash, fever	CLL, B-cell NHL

BCL-2-binding ligand to be discovered using a computer-based screening strategy.[47] HA14-1 induces apoptosis in several tumor cell lines and enhances their apoptotic responses to γ-radiation and novel anticancer agents, such as recombinant TRAIL, bortezomib, flavopiridol, imatinib and MAPK kinase inhibitors.[48-50] HA14-1 also overcomes intrinsic and acquired chemotherapy resistance caused by BCL-2 over-expression.[51] Skommer et al demonstrated synergy between HA14-1 and dexamethasone and doxorubicin, but not vincristine, in B-cell lymphoma.[52] They observed that for HA14-1 and doxorubicin, synergistic inhibition was achieved only when the BCL-2 inhibitor was administered 24 hours before the conventional chemotherapeutic drugs, while clear antagonism was seen when the two drugs were administered simultaneously. In combination with the cyclin-dependent kinase (CDK) inhibitor flavopiridol, HA14-1 exhib-ited a similar synergy in a multiple myeloma model, but in this instance, only when the BCL-2 inhibitor is administered after the CDK inhibitor.[50] While the biochemical basis for this synergy requires clarification, it is clear that schedule dependency will be an important feature of these agents and combinations with them.

BH3I-2 and Analogs

Small molecule inhibitors of the Bcl-x$_L$—BH3 domain interaction (BH3Is) disrupt in-teractions between BCL-X$_L$ and pro-apoptotic BCL-2 family proteins at low micromolar concentrations. BH3I-2′, an analog of BH3I-2, inhibits mitochondrial respiration, damages inner mitochondrial membrane and induces apoptosis via caspase-dependent and -independent mechanisms.[53,54] It also sensitizes TRAIL-induced apoptosis in leukemia cells.[48]

Antimycin and Analogs

The mitochondrial electron chain inhibitor Antimycin A, isolated from Streptomyces, is an antifungal compound that interacts with the hydrophobic BH3 binding groove of BCL-X$_L$. This interaction competitively inhibits the interaction with BH3 peptides binding to BCL-2, resulting in caspase-independent apoptosis in cells expressing high levels of BCL-X$_L$.[55-57] The compound is highly toxic in mouse models, with a single median lethal dose of 1 mg/kg, likely owing to the brad nonspecific effects the compound has on normal cells as well. Antimycin A analogs such as 2-methoxyantimycin inhibit the effects of BCL-X$_L$[58] and exhibit antitumor activities in preclinical models[59] and may show more selectivity in preclinical models.

Gossypol and Derivatives

Gossypol is an orally-available compound found in cottonseeds originally used as an herbal medicine in China.[60] The (-)-enantiomer ((-)-Gossypol, AT-101), binds to the BH3-binding grooves of BCL-2, BCL-X$_L$ and Mcl-1, displacing BH3 peptides with inhibitory concentrations 50% in the sub-micromolar range (IC$_{50}$).[61] (-)-Gossypol promotes an allosteric conformational change in BCL-2 and loss of mitochondrial membrane potential in a BAX/BAK-independent fashion.[62] (-)-Gossypol-induced apoptosis involves cytochrome c release from mitochondria and activation of several caspases.[63] (-)-Gossypol has also recently been demonstrated to improve the efficacy of cyclophosphamide-adriamycin-vincristine-prednisone (CHOP) and cyclophosphamide-rituximab based regimens in lymphoma xenograft models.[63,64] Paoluzzi et al[64] have shown the importance of a sequential administration of the BH3-mimeitc AT-101 and cyclophosphamide in in vitro and in vivo models of diffuse large B-cell lymphoma. In vitro data suggested that a pre-exposure to AT-101 for up to 48 hours was necessary to obtain synergism with the cyclophosphamide metabolite 4-HC. In vivo mouse studies also confirmed that a pre-exposure to AT-101 before administering rituximab and cyclophosphamide is beneficial in terms of tumor volume control (Fig. 1).

A Phase 2 open label trial is presently evaluating the safety and activity of AT-101 administered on one of two different dosing schedules in combination with Rituximab in patients with relapsed or refractory chronic lymphocytic leukemia (CLL).[65] Based on a previous report from Castro et al,[66] 12 patients received up to 3 months of AT-101 (30 mg daily for 3 out of every 4 weeks) with rituximab (375 mg/m^2 × 12 doses) on days 1, 3, 5, 8, 15, 22, 29, 31, 33, 40, 57, 59, 61. The results from a second cohort (n = 6) treated with intermittent, "pulse" AT-101, 80 mg/d on days 1-3 and 15-17 of each 28-day cycle, in combination with weekly rituximab, 375 mg/m^2/week were reported separately. Based on these data and the 6 patients treated with "pulse" AT-101, gastrointestinal (GI) toxicity was the most notable adverse effect for AT-101 on daily administration, was reduced with the pulse exposure. Interestingly, 2 of 6 patients on the intermittent schedule experienced Grade 1-2 GI toxicity and no patient experienced Grade 3-4 ileus, compared to 11 of 12 and 2 of 12 patients, who, respectively, experienced these symptoms on the in the daily schedule of AT-101. Apoptosis of CLL cells evaluated by flow-cytometry at the time of maximum (Cmax) AT-101 concentration was evident in 18-45% of cells in 4 of the 6 patients after a single 80 mg dose of AT-101. By comparison, apoptosis after a 30 mg AT-101 dose appeared lower and was detected in approximately 1-15% of cells. After 80 mg of AT-101, plasma concentrations up to 6.6 μM were observed compared with concentrations of approximately 0.8-1.8 μM after a 30 mg dose on the daily schedule. In the "pulse" AT-101 cohort partial responses (PR) were observed in 3 patients, while response was too early to be evaluated in the other 3 patients. Five of 12 patients experienced a PR on the continuous dosing schedule. Collectively, these data suggested that intermittent administration of AT-101 with a "pulse" dose regimen was associated with higher plasma concentrations and increased apoptosis, as well as reduced toxicity, when compared to the continuous daily schedule.

Given the reactivity and pharmacologic properties of the gossypol derivatives,[67] efforts have been focused on the synthesis of semi-synthetic analogs with improved pharmacologic properties. One of those derivatives, apogossypol was synthesized and characterized using a combination of molecular design approaches including molecular modeling, magnetic resonance (NMR)-based

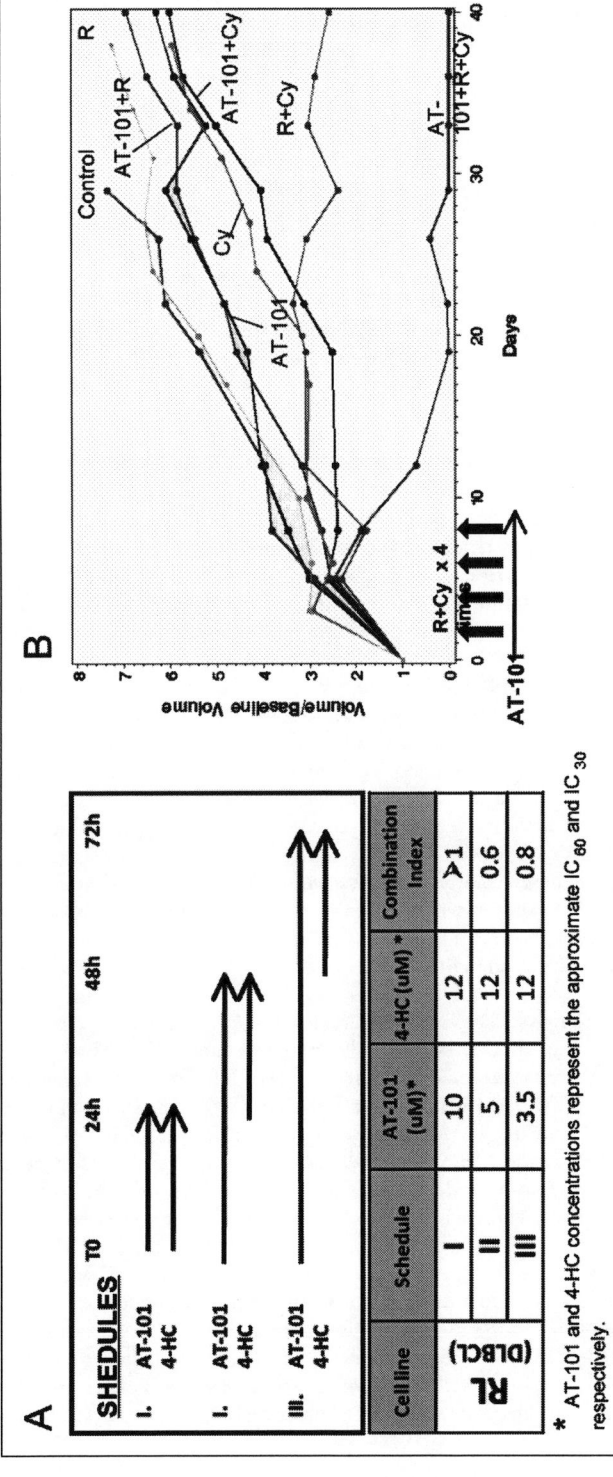

Figure 1. Diffuse Large-B-cell lymphoma model (RL) A) Cytotoxicity assay for AT-101 combined with 4-HC. Model of in vitro exposure to AT-101 and 4-HC; a pre-exposure to AT-101 for up to 48 hours before adding 4-HC for additional 24 hours (Schedules I and II) revealed a synergistic interaction of the combination (combination index less than 1) while a simultaneous exposure for 24 hours (schedule I) shows antagonism (combination index greater than (1). B) In vivo SCID–beige xenograft model for DLBCL (RL). The combination of oral AT-101 35 mg/kg/day for ten days regimen plus i.p. cyclophosphamide (Cy) and i.p. rituximab (R) in four administrations together on days 2, 4, 6, 8 showed significant tumor volume control compared to any other treatment group. The multiple comparison analysis showed the superiority of the triplet combination to each other group (p ≤ 0.0341). N = 7 in each group.[64]

structural analysis, fluorescence polarization assays and cell-based assays.[68] Apogossypol binds to and inhibits BCL-2, BCL-X$_L$, Mcl-1, BCL-2 and Bcl-B with high affinity, inducing apoptosis in tumor cell lines in the sub-micromolar range. Apogossypol was recently shown to have superior efficacy and less toxicity in transgenic mouse models of follicular lymphoma.[69]

Apogossypolone, another gossypol derivative that exhibits submicromolar binding affinity for BCL-2, MCL1 and BCL-X$_L$, has demonstrated activity in cell lines representing follicular lymphoma, mantle cell lymphoma, marginal zone lymphoma and chronic lymphocytic leukaemia. In a xenograft mouse model using a unique EBV-negative low grade lymphoma line (WSU-FSCCL) exhibiting both the t(14;18) and t(8;11) translocations, apogossypolone produced extended survival times, protecting mice from the bone-marrow infiltrating effects of the disease.[70]

TW37, another gossypol derivative with submicromolar affinities for BCL-2 and Mcl-1, was developed using computational screening and NMR spectroscopy focused on targeting discrete BCL-2 protein family members.[71] Preclinical studies have demonstrated that TW-37 is effective against non-Hodgkin's lymphoma and leukemia cell lines with little toxicity to normal peripheral blood lymphocytes.[72,73]

GX015-070 (Obatoclax Mesylate)

GX015-070 (Obatoclax mesylate) is an indole-derivative and a broad-spectrum inhibitor of pro-survival BCL-2 family proteins. It activates the mitochondrial apoptotic pathway by displacing BAK from Mcl-1 and BCL-X$_L$, upregulating BIM and inducing BAX and BAK conformational changes, mitochondrial depolarization and caspase activation.[74-76] The small molecule has been shown to synergize with the proteasome inhibitor bortezomib in mantle cell lymphoma (MCL), while not producing any significant cytotoxicity against peripheral blood mononuclear cells (PBMC) from healthy donors.[75] As both a single agent and in combination with melphalan, dexamethasone, or bortezomib, obatoclax effectively induces apoptosis in a variety of cell lines, including patient derived myeloma.[76] Recently obatoclax has also shown synergistic antileukemia effect when combined with the histone deacetylase inhibitor MGCD0103.[77]

O'Brien et al recently reported the results of a Phase I trial in patients with advanced CLL in which obatoclax was administered in doses ranging from 3.5 to 14 mg/m^2 as a 1-hour infusion and from 20 to 40 mg/m^2 as a 3 hour infusion every 3 weeks. Twenty-six patients were treated on this study.[78] Dose limiting reactions were primarily neurologic (somnolence, euphoria, ataxia) and associated with the infusion. The MTD was determined to be 28 mg/m^2 over 3 hours every 3 weeks. One patient (4%) achieved a partial response, while some patients with anemia (3/11) and thrombocytopenia (4/14) experienced improvements in hemoglobin and platelet counts.

Schimmer et al evaluated a prolonged infusion schedule of obatoclax in order to minimize toxicities while trying to maintaining clinical activity.[79] Forty-four patients with refractory hematologic malignancies received obatoclax from 7 to 28 mg/m^2 as a 24 hour infusion every other week or weekly. The principle histologies included: acute myelogenous leukemia (AML; 25), myelodysplastic syndromes (MDS; 14), acute lymphocytic leukemia (ALL; 1) and CLL (4). The most common adverse events included grade 1 and 2 CNS symptoms including somnolence (43%), dizziness (38%), euphoric mood (34%), fatigue (36%), gait disturbance (34%), diarrhea (30%), nausea and vomiting (31%). Interestingly, there were no DLTs, though given the adverse CNS events at the highest dose evaluated, it was determined that further escalation of the drug beyond 28 mg/m^2 dose would not be prudent. Obatoclax exhibited a favorable pharmacokinetic profile, rapidly achieving steady-state plasma levels, with the Cmax and AUC being proportional to the dose administered. One patient with treatment-related AML with a t(9;11)(p22;q23) translocation achieved a cytogenetic complete response (CR) with complete hematological recovery and transfusion independence by day 9 following the start of weekly 24 h infusions. This CR was sustained for 8 months. The authors concluded that obatoclax could be administered by prolonged infusions without producing additional cumulative toxicities. The dramatic response observed in a patient with AML with immediate recovery of peripheral blood counts supports the notion that the cytotoxic effects of obatoclax are specific to malignant cells while sparing normal bone marrow cells. Obatoclax also demonstrated activity in myelodisplasia as 3 of 14 patients experienced hematologic improvement and became transiently platelet or RBC transfusion independent.

Based on the data published by Galan-Perez et al demonstrating a synergy between obato-clax and the proteaseome inhibitor bortezomib, a combination Phase I study of obatoclax and bortezomib was launched.[80] Both obatoclax and bortezomib were administered on days 1, 4, 8 and 11 of a 21 day cycle, with up to 8 cycles of therapy being administered. Three patients received 30 mg of obatoclax and 1.0 mg/m^2 of bortezomib, 3 received 30 mg of obatoclax and 1.3 mg/m^2 of bortezomib and 6 received 45 mg of obatoclax and 1.3 mg/m^2 of bortezomib, which was the highest dose evaluated. Efficacy evaluation was performed every 2 cycles (approximately every 6 weeks). All patients had prior anthracyclines and rituximab and 5 had previously received bortezomib. The preliminary data based 9 patients, revealed an overall incidence of thrombocytopenia in more than one-quarter of the patients. Other adverse toxicities included abdominal distension, abdominal pain, constipation, diarrhea, nausea, fatigue, weight loss, dehydration, neuropathy, somnolence, euphoric mood, cough and rash. The most common grade 3 and grade 4 adverse event was thrombocytopenia (22%). While somnolence and euphoric mood resolved soon after the infusion ended, there were no grade 3 or 4 CNS toxicities. Obatoclax mesylate at 45 mg and bortezomib at a dose of 1.3 mg/m^2 was determined to be the combination dosage for further Phase 2 study. Based on investigator reported assessments, 2 patients in the obatoclax 30 mg/bortezomib 1.0 mg/m^2 dosage group and 1 patient in the obatoclax 30 mg/bortezomib 1.3 mg/m^2 dosage group achieved a CR/CRu. Two of these patients had prior high dose therapy with autologous stem cell transplants, while the third had prior bortezomib. The authors conclude that obatoclax and bortezomib (45 mg and 1.3 mg/m^2, respectively) administered on days 1, 4, 8 and 11 of a 21 day cycle have acceptable tolerability, with an albeit early signal of activity.

ABT-737 and ABT-263

ABT-737 is a synthetic small-molecule which is now considered to be the most potent and specific BCL-2/BCL-X$_L$ inhibitor discovered so far.[81] ABT-737 has extremely high affinity for BCL-X$_L$, BCL-2 and BCL-2, with a dissociation constant (Ki) below 1 nM for each protein and binds poorly to Mcl-1 and A1. ABT-737 was rationally designed to be a BH3 only mimetic designed after the BH3 protein based on the structural properties of Bad. While it does not bind to BAX, it is known to disrupt the complex of BAX and BCL-2 triggering a conformational alteration of BAX. ABT-737 also displaces BH3-only proteins such as BIM from its binding partners. Interestingly, the effects of ABT-737 are completely abrogated in BAX and BAK deficient cells.[82] This strict dependence on BAX and BAK distinguishes ABT-737 from other small-molecule BCL-2/BCL-X$_L$ inhibitors and suggests it function as a strict authentic BH3 mimetic. As a single agent, ABT-737 is active against several lymphoid malignancies including follicular, diffuse large B-cell and mantle cell lymphoma, chronic lymphocytic leukemia, acute lymphocytic and myeloid leukemia and multiple myeloma with IC50s in the nanomolar or low micromolar range.[81,83-90] Apoptosis induced by ABT-737 is associated with cytochrome *c* release from the mitochondria and activation of caspases.[81,82,85,88,91]

ABT-737 exhibits marked synergy when combined with γ-irradiation as well as a variety of anticancer agents including etoposide, doxorubicin, cisplatin, melphalan, Ara-C, paclitaxel, vincristine, dexamethasone, thalidomide and bortezomib.[81,82,88,92] It has been shown to overcomes resistance to the Bcr-Abl inhibitors imatinib and INNO-406 in leukemia cells carrying the Bcr/Abl translocation[93] and enhances the anticancer effects of several investigational agents, such as the cyclin-dependent kinase (CDK) inhibitor roscovitine, the MDM2 inhibitor Nutlin-3a and MEK inhibitors.[88,94,95] ABT-737 treatment in general is well tolerated by normal hematopoietic cells and bone marrow cells.[81,85]

The antitumor activities of ABT-737 have been characterized in several animal models. ABT-737 suppresses tumor growth in xenograft mouse models of myeloma and mantle cell lymphoma.[90,96] In models of aggressive leukemia driven by Raf-transformed myeloid cells, it suppressed tumor growth by about 50%, significantly extending survival of the treated mice.[85] In different studies, ABT-737 has shown to be well tolerated without any significant weight loss (<5%), but does cause reduction in platelet and lymphocyte counts.[81] Several groups have reported that Mcl-1 expression plays a major role in causing ABT-737 resistance. Leukemia and small cell lung cancer (SCLC) cells expressing relatively higher levels of BCL-2 and BCL-X$_L$ and lower levels of Mcl-1 were found to be sensitive to ABT-737.[83,97] Conversely, those expressing high levels of Mcl-1 were resistant to ABT-737.[82,85] Unbiased genomic analysis and siRNA library screening also identified Mcl-1 and Noxa as modulators of ABT-737 sensitivity.[98,99] Remarkably, combinations of ABT-737 with agents that decrease Mcl-1 expression, such as roscovitine, cycloheximide or arsenic trioxide, markedly augmented the effects of ABT-737 on human leukemia and SCLC cell lines.[94,100] These studies not only validated the specificity and molecular mechanisms of ABT-737, but also provided a rationale for targeting Mcl-1 for improving its therapeutic efficacy. Based on these collective data, ABT-737 is likely to be more efficacious as a single agent for those tumors in which Mcl-1 expression is low, absent, or inactivated, such as follicular lymphoma and chronic lymphocytic leukemia. Conversely, for those tumors in which Mcl-1 is the predominant survival protein, ABT-737 is unlikely to be effective as a single agent, but may compliment other therapeutic agents that down-regulate Mcl-1.[82,85,94] Combination therapies using genotoxic agents and ABT-737 could be particularly effective, as many genotoxic drugs induce Mcl-1 degradation which should greatly potentiate the effects of ABT-737. Furthermore, the rapid turnover of Mcl-1 mRNA and protein provide the rationale for combining ABT-737 with inhibitors of transcription or translation such as CDK inhibitors and multi-kinase inhibitors. Mcl-1 degradation is regulated by the ubiquitin E3 ligase Mule,[94] which may be manipulated to enhance the therapeutic effects of ABT-737. Although ABT-737 seems to be well tolerated in animals, its ability to inhibit several pro-survival proteins in normal cells might still be a concern for causing adverse effects. For example, ABT-737 causes dose-dependent acute thrombocytopenia by reducing the number of circulating platelets whose turnover is regulated by apoptosis. Platelets are particularly sensitive to ABT-737, perhaps because of BAK-dependent apoptosis normally constrained by BCL-X$_L$ in these cells. The process of platelet activation and senescence in vivo is associated with processes similar to those observed during apoptosis in nucleated cells, including loss of mitochondrial membrane potential, caspase activation, phosphatidylserine externalization and cell shrinkage. Recent data suggest that ABT-737 induces an apoptosis-like response in platelets that is distinct from platelet activation and results in enhanced clearance in vivo by the reticuloendothelial system.[101]

ABT-263 is structurally similar to ABT-737 and is the derivative that has moved into the clinic. It is a second generation small molecule BCL-2 family protein inhibitor that binds with high affinity ($K_i \leq 1$ nM) to multiple anti-apoptotic BCL-2 family proteins including BCL-X$_L$, BCL-2, BCL-2 and Bcl-B has recently shown potent mechanism-based cytotoxicity ($EC_{50} \leq 1$ mM) against human tumor cell lines derived from lymphoid malignancies and SCLC. ABT-263 is presently under Phase 1/2 study in patients with lymphoid malignancies, SCLC/solid tumors and CLL. In a Phase 1/2 multicenter study, ABT-263 is being administered orally for 14 consecutive days of a 21 day cycle.[102] Pharmacokinetic (PK) assessment was performed at study initiation and on day 14 of drug dosing. Presently, 17 subjects have been enrolled in the lymphoma study. Three subjects completed each of the 10, 20, 40 and 80 mg cohorts enrolled thus far. A Grade 3 DLT (upper respiratory infection) occurred in the 160 mg cohort, which was then expanded to 6 patients. Two patients with bulky SLL/CLL in the 40 and 160 mg cohorts experienced 95% and 64% tumor reductions after cycles 4 and 2, respectively and continue on treatment. The PK profile of ABT-263 is linear between the 10 mg and 160 mg dose levels. The average terminal half-life of ABT-263 has ranged from 14 to 25 hours across all dose levels. ABT-263 reduced the platelet level in a dose-dependent manner. One subject experienced a Grade 3 thrombocytopenia postdosing with 80 mg, resolving 8 hours later. The decreased platelet count did not result in any medical sequelae.

Antisense Strategies

Oblimersen sodium (BCL-2 antisense oligonucleotide; G3139; Genasense) is a single-stranded 18 mer oligodeoxyribonucleotide, complimentary to the first six codons of the human bcl-2 reading frame. Preclinical studies of oblimersen on the BCL-2 overexpressing lymphoma cell lines DoHH2 and SU-DHL-4 in vitro have reported reduction in BCL-2 protein in treated cells.[103] To date, a number of in vitro studies have shown synergistic enhancement of tumour cell killing when BCL-2 antisense was combined with other standard therapies, including alkylating agents and proteasome inhibitors.[104,105] Combination of oblimersen with a CD20 monoclonal antibody (rituximab) in severe combined immunodeficiency (SCID)/human lymphoma xenograft models of lymphoma produced a significant improvement in survival.[106] Based on these data, Pro et al[107] conducted a Phase II study of oblimersen sodium and rituximab in patient with recurrent B-cell NHL to determine the efficacy and safety of this combination.

In this study, oblimersen was administered as a continuous intravenous infusion at a daily dose of 3 mg/kg/d for 7 days on alternate weeks for 3 weeks. Rituximab was given at a weekly dose of 375 mg/m^2 for six doses. Patients with stable disease or objective response were allowed to receive a second course of treatment. The overall response rate (ORR) was 42% with 10 complete responses (CR) and eight partial responses (PR). Twelve (28%) patients achieved a minimal response or stable disease. Among the 20 patients with follicular lymphoma the ORR was 60% (eight CR, four PR), despite three of the responders being refractory to prior treatment with rituximab and two of the responses having failed prior autologous stem cell transplant. Interestingly, the median duration of response was 12 months. Most toxicities were felt to be relatively low grade and reversible.

Recently O'Brien et al,[108] presented the results of a 5-year follow-up on patients with relapsed/refractory CLL treated with standard chemotherapy (fludarabine/cyclophosphamide) with or without oblimersen in a randomized Phase III trial. Two hundred forty-one patients were stratified according to response to prior fludarabine therapy (relapsed sensitive versus refractory); number of prior regimens (1, 2, 3 or more) and duration of response to last therapy (more or less than 6 months). Treatment included up to six 28-day cycles of oblimersen at 3 mg/kg/d as a 7-day continuous IV infusion and fludarabine at 25 mg/m^2/day IV followed by cyclophosphamide at 250 mg/m^2/d IV on days 5, 6 and 7, compared to the fludarabine and cyclophosphamide regimen alone on days 1, 2 and 3. The complete remission rate was 17% in the oblimersen combination arm and 7% in the arm without oblimersen (p = 0.025). Only 7 subjects were lost to 5-year follow-up. Univariate analyses confirmed the prognostic value of the trial stratification factors for survival. In addition, baseline beta-2 microglobulin (less or more than 4 mg/L) and baseline serum LDH (less or more than ULN) were highly prognostic (p < 0.0001). Multivariate analyses showed that the number of prior regimens, age (less or more than 65 years), baseline beta-2 microglobulin and baseline serum LDH were prognostic for survival. Among all prognostic factors examined in a multivariate analyses, a significant interaction with treatment was detected only for fludarabine sensitivity. Maximum benefit with oblimersen was observed in fludarabine-sensitive patients, that is, those who had a partial response or better for ≥6 months after their last prior fludarabine-containing therapy who then relapsed. Disease-related findings, prognostic factors and prior treatments among fludarabine-sensitive patients were similar in the 2 treatment arms. The proportion of fludarabine-sensitive patients with CR was significantly greater in the oblimersen based arm (25%) than in the fludarabine/cyclophosphamide arm (6%; p = 0.016). With 5-year follow-up, there was a statistically significant survival benefit in the oblimersen arm among fludarabine-sensitive pts (hazard ratio = 0.50; p = 0.004). The authors concluded that in the setting of relapsed/refractory CLL, the number of prior regimens, age, beta-2 microglobulin and serum LDH were prognostic for survival based on the multivariate analyses. Fludarabine sensitivity was predictive of benefit in the oblimersen/fludarabine/cyclophosphamide regimen, producing a 50% reduction in the risk of death among this population of patients.

Targeting TRAIL

The death receptors of the TNF super-family represent potential targets for promoting apoptosis in cancer. As death receptor-mediated apoptosis is thought to be independent on p53, cancers with inactivating p53 mutations might be susceptible to treatment using this approach.

Agonistic Antibodies

Efforts to engage the extrinsic pathways mediated by TRAIL have been facilitated by the development of agonistic antibodies targeting DR4[109] and DR5.[110,111] These antibodies have been shown to induce apoptosis in cancer cells but not in normal cells and appear to at least slow the growth of tumors in xenograft tumor models with little apparent systemic toxicity. Antibodies have the advantage of having a relatively long half-life and can invoke alternative mechanisms for cell killing through antibody-dependent cellular cytotoxicity (ADCC) and complement-dependent cytotoxicity (CDC) mechanisms mediated by their Fc portion. Presently, several TRAIL receptor agonists are being evaluated in Phase I and/or II clinical trials, including antibodies targeting DR4 (mapatumumab) and DR5 (apomab and CS-1008). A Phase 2 study with mapatumumab in patients with relapsed or refractory non-Hodgkin's Lymphoma reported three responses in 14 patients with follicular lymphoma including one complete response.[112] A Phase II study with mapatumumab plus bortezomib is presently ongoing in patients with relapsed/refractory multiple myeloma. Another Phase II trial exploring the combination or apomab plus rituximab in rituximab resistant NHL is also ongoing.

Ligand Based Receptor Agonism

Another approach for pharmacologically triggering TRAIL receptors involves the use of soluble truncated versions of TRAIL that contain the extracellular domain. In preclinical studies, recombinant TRAIL (rhApo2L/TRAIL) induced apoptosis in various cancer cell lines, including those with p53 mutations, without affecting normal cells. In addition, chemotherapeutic drugs and histone deacetylase inhibitors were shown to augment the apoptotic activity of TRAIL agonists.[113,114] TRAIL also displayed antitumor activity in vivo in mouse models of multiple myeloma, when given as a single agent.[115] Although some recombinant forms of TRAIL have been shown to be toxic to hepatocytes and other normal cells, these effects are thought to be related to the particular recombinant forms of the protein rather than TRAIL itself.[116,117] Safety evaluations in primates with TRAIL that did not contain extraneous amino-acid residues showed no toxicities related to TRAIL exposure.[118] A Phase Ib study of rhApo2L/TRAIL plus rituximab in relapsed, low-grade NHL has been completed. The combination seemed to be well tolerated; with few exceptions, adverse events were mild to moderate and no maximally tolerated dose of rhApo2L/TRAIL with rituximab was reached. Two patients experienced a complete response, one a partial response and two patients a stable disease.[119]

BCL-2 Proteins and Non-Apoptotic Cell Death

BCL-2 has also been linked to non-apoptotic cell death as well.[120,121] Both BCL-2 and BCL-X_L can bind the tumour suppressor beclin 1 (BECN1), which also contains a BH3 domain. This interaction has been shown to inhibit autophagy, a process important for the degradation of bulk cytoplasm, long-lived proteins and entire organelles. Release of this interaction by competition with a BH3-only protein (or a small-molecule mimetic) has been found to stimulate autophagy.[122] The role of autophagy in disease pathogenesis has only recently begun to be appreciated. Unlike the ubiquitin-proteasome pathway that is responsible for degradation of short-lived cellular proteins, autophagy is responsible for the degradation of long-lived proteins. Furthermore, autophagy is responsible for promoting cellular survival under conditions of nutrient deprivation. However, uncontrolled and prolonged autophagy will also lead to cell death by degrading essential survival proteins or by degrading inhibitors of programmed cell death.[123] This observation has generated interest in the potential for therapeutic induction of autophagy in malignant tissues. While regulation of autophagy remains poorly understood, there is growing evidence that the mammalian target of rapamycin (mTOR), extracellular signal-regulated kinases (ERK) and several members of the bcl-2 protein family play important regulatory roles in this process.[121] The BCL-2 family exerts cell cycle effects in addition to regulating apoptosis. BCL-2 and Bcl-x_L upregulate p27 and promote arrest on the G(0) phase of the cell cycle.[124] Recently, Cui et al[125] explored whether autophagy is involved in BCL-2 and Bcl-x(L)-mediated cell cycle arrest and found that autophagy was activated, but not required, for G(0) arrest. They discovered that the cell cycle function of BCL-2 and Bcl-x(L) was dependent on BAX and BAK and in BAX(-/-) BAK(-/-) double knockout cells, features of G(0) quiescence were already present and p27 was constitutively elevated. The authors concluded that a physiological role of BAX and BAK may be the suppression of autophagy. Yazbeck et al[126] have recently shown how the mTOR inhibitor temsirolimus induces autophagy and synergizes with the histon deacetylase inhibitor vorinostat in mantle cell lymphoma.

While clearly only in its infancy, there is no doubt that the further clarification of this important survival pathway will open new therapeutic opportunities for the treatment of the lymphproliferative malignancies.

Conclusion

Targeting cellular death pathways including apoptosis is a promising strategy for cancer drug discovery. The fruits of over three decades of research on BCL-2-family proteins have yielded new strategies for therapeutic intervention, some of which have advanced to clinical studies in patients. The most advanced of these candidate therapeutics is an antisense oligonucleotide targeting BCL-2 mRNA (oblimersen sodium), which has shown promising activity for chronic lymphocytic leukemia in randomized Phase 3 clinical trials. Antibodies and recombinant TRAIL agonists that target DR4 and CD5 are also in clinical trials for patients with both solid tumors and hematologic malignancies. Several small-molecule antagonists of anti-apoptotic BCL-2-family proteins have been described that bind the same pocket occupied by pro-apoptotic BH3 domains. These compounds have different potencies and a variable spectrum of activity against the six anti-apoptotic members. Bcl2-family inhibitors are showing impressive preclinical efficacy in animal models and are moving rapidly towards Phase I and II clinical trials. A common hurdle that all of these small-molecule-based therapies have had to overcome is the difficulty in targeting a discrete protein–protein interaction. To aid in this process, the three-dimensional structures of the natural ligands complexed to their protein targets were determined and used for the design of very precise inhibitors. Linked fragment-based approaches and parallel synthesis have also been used. The knowledge gained from developing these approaches might prove useful in the future for designing inhibitors for other difficult targets. For these pro-apoptotic molecules to succeed they will need to be thoroughly evaluated in the clinic, with close attention to both the pharmacodynamic and pharmacokinetics aspects of drug development. Since lymphoma is such a heterogeneous disease, it will be crucial to appreciate the spectrum of activity of these agents across the different sub-types of NHL< and to understand the differences in 'survival biology' within these distinct NHL histologies. While there are high hopes for these agents, it may be that their greatest utility will come in lowering the apoptotic threshold of any given lymphoma, sensitizing to the pro-apoptotic effects of other chemotherapeutic agents. Given the need to understand molecular pharmacology in detail, appropriate preclinical studies will need to identify the optimal strategies for combining these agents, with an emphasis on the importance of dose and schedule dependency.

References

1. Reed JC. BCL-2-family proteins and hematologic malignancies: history and future prospects. Blood 2008; 111(7):3322-30. Review. Erratum in: Blood 2008; 112(2):452.
2. Vaux DL, Cory S, Adams JM. BCL-2 gene promotes haemopoietic cell survival and cooperates with c-Myc to immortalize preB-cells. Nature 1988; 335:440-442.
3. Kerr JFR, Wyllie AH, Currie AR. Apoptosis: a basic biological phenomenon with wide-ranging implications in tissue kinetics. Br J Cancer 1972; 26:239-257.

4. Vaux DL, Weissman IL, Kim SK. Prevention of programmed cell death in Caenorhabditis elegans by human BCL-2. Science 1992; 258:1955-1957.
5. Yuan J, Shaham S, Ledoux S et al. The C. elegans cell death gene ced-3 encodes a protein similar to mammalian interleukin-1β-converting enzyme. Cell 1993; 75:641-652.
6. Li P, Nijhawan D, Budihardjo I et al. Cytochrome c and dATP-dependent formation of Apaf-1/caspase-9 complex initiates an apoptotic protease cascade. Cell 1997; 91:479-489.
7. Oltvai ZN, Milliman CL, Korsmeyer SJ. BCL-2 heterodimerizes in vivo with a conserved homolog, Bax, that accelerates programmed cell death. Cell 1993; 74:609-619.
8. Reed JC. Apoptosis-based therapies. Nat Rev Drug Discov 2002; 1(2):111-21. Review.
9. Bredesen DE, Rao RV, Mehlen P. Cell death in the nervous system. Nature 2006; 443(7113):796-802. Review.
10. Waterhouse NJ, Goldstein JC, von Ahsen O et al. Cytochrome c maintains mitochondrial transmembrane potential and ATP generation after outer mitochondrial membrane permeabilization during the apoptotic process. J Cell Biol 2001; 153(2):319-28.
11. Pattingre S, Tassa A, Qu X et al. BCL-2 antiapoptotic proteins inhibit Beclin 1-dependent autophagy, Cell 2005; 122:927-939.
12. Krajewski S, Tanaka S, Takayama S et al. Investigation of the subcellular distribution of the bcl-2 onco-protein: residence in the nuclear envelope, endoplasmic reticulum and outer mitochondrial membranes. Cancer Res 1993; 53(19):4701-14.
13. Ron D. Cell biology. Stressed cells cope with protein overload. Science 2006; 313(5783):52-3.
14. Adams JM, Cory S. The BCL-2 apoptotic switch in cancer development and therapy. Oncogene 2007; 26:1324-1337.
15. Youle RJ, Strasser A. The BCL-2 protein family: opposing activities that mediate cell death. Nat Rev Mol Cell Biol 2008; 9(1):47-59. Review.
16. Wei MC, Zong WX, Cheng EH et al. Proapoptotic BAX and BAK: a requisite gateway to mitochondrial dysfunction and death. Science 2001; 292(5517):727-30.
17. Huang DC, Strasser A. BH3-only proteins-essential initiators of apoptotic cell death. Cell 2000; 103:839-842.
18. Sattler M, Liang H, Nettesheim D et al. Structure of Bcl-xL-Bak peptide complex: recognition between regulators of apoptosis. Science 1997; 275:983-986.
19. Petros AM, Nettesheim DG, Wang Y et al. Rationale for Bcl-xL/Bad peptide complex formation from structure, mutagenesis and biophysical studies. Protein Sci 2000; 9:2528-2534.
20. Cheng EH, Wei MC, Weiler S et al. BCL-2, BCL-X(L) sequester BH3 domain-only molecules preventing BAX- and BAK-mediated mitochondrial apoptosis. Mol Cell 2001; 8:705-711.
21. Danial NN, Korsmeyer SJ. Cell death. Critical control points. Cell 2004; 116:205-219.
22. Zong WX, Lindsten T, Ross AJ et al. BH3-only proteins that bind pro-survival BCL-2 family members fail to induce apoptosis in the absence of Bax and Bak. Genes Dev 2001; 15:1481-1486.
23. Green DR, Kroemer G. The pathophysiology of mitochondrial cell death. Science 2004; 305:626-629.
24. Wang X. The expanding role of mitochondria in apoptosis. Genes Dev 2001; 15:2922-2933.
25. Lessene G, Czabotar PE, Colman PM. BCL-2 family antagonists for cancer therapy. Nat Rev Drug Discov 2008; 7(12):989-1000. Review.
26. Chen L, Willis SN, Wei A et al. Differential targeting of prosurvival BCL-2 proteins by their BH3-only ligands allows complementary apoptotic function. Mol Cell 2005; 17:393-403.
27. Kuwana T, Bouchier-Hayes L, Chipuk JE et al. BH3 domains of BH3-only proteins differentially regulate Bax-mediated mitochondrial membrane permeabilization both directly and indirectly. Mol Cell 2005; 17(4):525-35.
28. Letai A, Bassik MC, Walensky LD et al. Distinct BH3 domains either sensitize or activate mitochondrial apoptosis, serving as prototype cancer therapeutics. Cancer Cell 2002; 2:183-192.
29. Cartron PF, Gallenne T, Bougras G et al. The first alpha helix of Bax plays a necessary role in its ligand-induced activation by the BH3-only proteins Bid and PUMA. Mol Cell 2004; 16(5):807-18.
30. Certo M, Del Gaizo Moore V, Nishino M et al. Mitochondria primed by death signals determine cellular addiction to antiapoptotic BCL-2 family members. Cancer Cell 2006; 9:351-365.
31. Kim H, Rafiuddin-Shah M, Tu HC et al. Hierarchical regulation of mitochondrion-dependent apoptosis by BCL-2 subfamilies. Nat Cell Biol 2006; 8:1348-1358.
32. Willis SN, Fletcher JI, Kaufmann T et al. Apoptosis initiated when BH3 ligands engage multiple BCL-2 homologs, not Bax or Bak. Science 2007; 315:856-859.
33. Oda E, Ohki R, Murasawa H et al. Noxa, a BH3-only member of the BCL-2 family and candidate mediator of p53-induced apoptosis. Science 2000; 288:1053-1058.
34. Nakano K, Vousden KH. PUMA, a novel proapoptotic gene, is induced by p53. Mol Cell 2001; 7:683-694.

35. Strasser A. The role of BH3-only proteins in the immune system. Nature Rev Immunol 2005; 5:189-200.
36. Puthalakath H, O'Reilly LA, Gunn P et al. ER stress triggers apoptosis by activating BH3-only protein Bim. Cell 2007; 129:1337-1349.
37. Kelley SK, Harris LA, Xie D et al. Preclinical studies to predict the disposition of Apo2L/tumor necrosis factor-related apoptosis-inducing ligand in humans: characterization of in vivo efficacy, pharmacokinetics and safety. J Pharmacol Exp Ther 2001; 299:31-38.
38. Motoyama N, Wang F, Roth KA et al. Massive cell death of immature hematopoietic cells and neurons in Bcl-x-deficient mice. Science 1995; 267:1506-1510.
39. Veis DJ, Sorenson CM, Shutter JR et al. J. BCL-2-deficient mice demonstrate fulminant lymphoid apoptosis, polycystic kidneys and hypopigmented hair. Cell 1993; 75:229-240.
40. Mason KD, Carpinelli MR, Fletcher JI et al. Programmed anuclear cell death delimits platelet life span. Cell 2007; 128:1173-1186.
41. Rinkenberger JL, Horning S, Klocke B et al. Mcl-1 deficiency results in peri-implantation embryonic lethality. Genes Dev 2000; 14:23-27.
42. Ross AJ, Waymire KG, Moss JE et al. Testicular degeneration in Bclw-deficient mice. Nature Genet 1998; 18:251-6.
43. Walensky LD, Kung AL, Escher I et al. Activation of apoptosis in vivo by a hydrocarbon-stapled BH3 helix. Science 2004; 305:1466-1470.
44. Zeitlin BD, Zeitlin IJ, Nör JE. Expanding circle of inhibition: small-molecule inhibitors of BCL-2 as anticancer cell and antiangiogenic agents. J Clin Oncol 2008; 26(25):4180-8. Review.
45. An J, Chen Y, Huang Z. Critical upstream signals of cytochrome C release induced by a novel BCL-2 inhibitor. J Biol Chem 2004; 279(18):19133-40.
46. Milanesi E, Costantini P, Gambalunga A et al. The mitochondrial effects of small organic ligands of BCL-2: sensitization of BCL-2-overexpressing cells to apoptosis by a pyrimidine-2,4,6-trione derivative. J Biol Chem 2006; 281(15):10066-72.
47. Wang JL, Liu D, Zhang ZJ et al. Structure-based discovery of an organic compound that binds bcl-2 protein and induces apoptosis of tumor cells. Proc Natl Acad Sci USA 2000; 97:7124-7129.
48. Hao JH, Yu M, Liu FT et al. BCL-2 inhibitors sensitize tumor necrosis factor-related apoptosis-inducing ligand-induced apoptosis by uncoupling of mitochondrial respiration in human leukemic CEM cells. Cancer Res 2004; 64(10):3607-16.
49. Milella M, Estrov Z, Kornblau SM et al. Synergistic induction of apoptosis by simultaneous disruption of the BCL-2 and MEK/MAPK pathways in acute myelogenous leukemia. Blood 2002; 99(9):3461-4.
50. Pei XY, Dai Y, Grant S. The small-molecule bcl-2 inhibitor HA14-1 interacts synergistically with flavopiridol to induce mitochondrial injury and apoptosis in human myeloma cells through a free radical-dependent and jun NH2-terminal kinase-dependent mechanism. Mol Cancer Ther 2004; 3:1513-1524.
51. Oliver L, Mahé B, Grée R et al. HA14-1, a small molecule inhibitor of BCL-2, bypasses chemoresistance in leukaemia cells. Leuk Res 2007; 31(6):859-63. Epub 2007.
52. Skommer J, Wlodkowic D, Matto M et al. HA14-1, a small molecule bcl-2 antagonist, induces apoptosis and modulates action of selected anticancer drugs in follicular lymphoma B-cells. Leuk Res 2006; 30:322-331.
53. Degterev A, Lugovskoy A, Cardone M et al. Identification of small-molecule inhibitors of interaction between the BH3 domain and Bcl-xL. Nat Cell Biol 2001; 3(2):173-82.
54. Feng WY, Liu FT, Patwari Y et al. BH3-domain mimetic compound BH3I-2′ induces rapid damage to the inner mitochondrial membrane prior to the cytochrome c release from mitochondria. Br J Haematol 2003; 121(2):332-40.
55. Tzung SP, Kim KM, Basañez G et al. Antimycin A mimics a cell-death-inducing BCL-2 homology domain 3. Nat Cell Biol 2001; 3(2):183-91.
56. Manion MK, O'Neill JW, Giedt CD et al. BCL-XL mutations suppress cellular sensitivity to antimycin A. J Biol Chem 2004; 279:2159-2165.
57. Kim KM, Giedt CD, Basanez G et al. Biophysical characterization of recombinant human bcl-2 and its interactions with an inhibitory ligand, antimycin A. Biochemistry 2001; 40:4911-4922.
58. Schwartz PS, Manion MK, Emerson CB et al. 2-Methoxy antimycin reveals a unique mechanism for Bcl-x(L) inhibition. Mol Cancer Ther 2007; 6(7):2073-80.
59. Wang H, Li M, Rhie JK et al. Preclinical pharmacology of 2-methoxyantimycin A compounds as novel antitumor agents. Cancer Chemother Pharmacol 2005; 56(3):291-8.
60. Pellecchia M, Reed JC. Inhibition of anti-apoptotic BCL-2 family proteins by natural polyphenols: new avenues for cancer chemoprevention and chemotherapy. Curr Pharm Des 2004; 10(12):1387-98. Review.

61. Kitada S, Leone M, Sareth S et al. Discovery, characterization and structure-activity relationships studies of proapoptotic polyphenols targeting B-cell lymphocyte/leukemia-2 proteins. J Med Chem 2003; 46(20):4259-64.
62. Lei X, Chen Y, Du G et al. Gossypol induces Bax/Bak-independent activation of apoptosis and cytochrome c release via a conformational change in BCL-2. FASEB J 2006; 20(12):2147-9.
63. Mohammad RM, Wang S, Aboukameel A et al. Preclinical studies of a nonpeptidic small-molecule inhibitor of BCL-2 and Bcl-X(L) [(-)-gossypol] against diffuse large cell lymphoma. Mol Cancer Ther 2005; 4(1):13-21.
64. Paoluzzi L, Gonen M, Gardner JR et al. Targeting BCL-2 family members with the BH3 mimetic AT-101 markedly enhances the therapeutic effects of chemotherapeutic agents in in vitro and in vivo models of B-cell lymphoma. Blood 2008; 111(11):5350-8.
65. Castro JE, Loria JO, Aguillon AR et al. A phase II, open label study of AT-101 in combination with rituximab in patients with relapsed or refractory chronic lymphocytic leukemia. Evaluation of two dose regimens. Blood (ASH Annual Meeting Abstracts) 2007; 110:3119.
66. Castro JE, Loria JO, Aguillon AR et al. A phase II, open label study of AT-101 in combination with rituximab in patients with relapsed or refractory chronic lymphocytic leukemia. Blood (ASH Annual Meeting Abstracts) 2006; 108:2838.
67. Van Poznak C, Seidman AD, Reidenberg MM et al. Oral gossypol in the treatment of patients with refractory metastatic breast cancer: a phase I/II clinical trial. Breast Cancer Res Treat 2001; 66(3):239-48.
68. Becattini B, Kitada S, Leone M et al. Rational design and real time, in-cell detection of the proapoptotic activity of a novel compound targeting Bcl-X(L). Chem Biol 2004; 11(3):389-95.
69. Kitada S, Kress CL, Krajewska M et al. BCL-2 antagonist apogossypol (NSC736630) displays single-agent activity in BCL-2-transgenic mice and has superior efficacy with less toxicity compared with gossypol (NSC19048). Blood 2008; 111(6):3211-9.
70. Arnold AA, Aboukameel A, Chen J et al. Preclinical studies of Apogossypolone: a new nonpeptidic pan small-molecule inhibitor of BCL-2, BCL-XL and Mcl-1 proteins in Follicular Small Cleaved Cell Lymphoma model. Mol Cancer 2008; 7:20.
71. Wang G, Nikolovska-Coleska Z, Yang CY et al. Structure-based design of potent small-molecule inhibitors of anti-apoptotic BCL-2 proteins. J Med Chem 2006; 49(21):6139-42.
72. Mohammad RM, Goustin AS, Aboukameel A et al. Preclinical studies of TW-37, a new nonpeptidic small-molecule inhibitor of BCL-2, in diffuse large cell lymphoma xenograft model reveal drug action on both BCL-2 and Mcl-1. Clin Cancer Res 2007; 13(7):2226-35.
73. Mohammad RM, Sun Y, Wang S et al. Evaluation of TW-37, a pan BCL-2 proteins small-molecule inhibitor, against spectrum of human B-Cell lines and patient-derived samples. Blood (ASH Annual Meeting Abstracts) 2007; 110:4521.
74. Campàs C, Cosialls AM, Barragán M et al. BCL-2 inhibitors induce apoptosis in chronic lymphocytic leukemia cells. Exp Hematol 2006; 34(12):1663-9.
75. Pérez-Galán P, Roué G, Villamor N et al. The BH3-mimetic GX15-070 synergizes with bortezomib in mantle cell lymphoma by enhancing Noxa-mediated activation of Bak. Blood 2007; 109(10):4441-9. Epub 2007.
76. Trudel S, Li ZH, Rauw J et al. Preclinical studies of the pan-Bcl inhibitor obatoclax (GX015-070) in multiple myeloma. Blood 2007; 109(12):5430-8.
77. Wei Y, Kadia T, Tong W et al. The combination of a histone deacetylase (HDAC) inhibitor with the BCL-2 inhibitor GX15-070 has synergistic antileukemia effect by inducing both apoptotic and autophagic pathways. Blood (ASH Annual Meeting Abstracts) 2008; 112:1633.
78. O'Brien SM, Claxton DF, Crump M et al. Phase I study of obatoclax mesylate (GX15-070), a small molecule pan-BCL-2 family antagonist, in patients with advanced chronic lymphocytic leukemia. Blood 2009; 113(2):299-305. Epub 2008.
79. Schimmer AD, O'Brien S, Kantarjian H et al. A phase I study of the pan bcl-2 family inhibitor obatoclax mesylate in patients with advanced hematologic malignancies. Clin Cancer Res 2008; 14(24):8295-301.
80. Goy A, Ford P, Feldman T et al. A phase 1 trial of the pan BCL-2 family inhibitor obatoclax mesylate (GX15-070) in combination with bortezomib in patients with relapsed/refractory mantle cell lymphoma. Blood (ASH Annual Meeting Abstracts) 2007; 110:2569.
81. Oltersdorf T, Elmore SW, Shoemaker AR et al. An inhibitor of BCL-2 family proteins induces regression of solid tumours. Nature 2005; 435(7042):677-81.
82. van Delft MF, Wei AH, Mason KD et al. The BH3 mimetic ABT-737 targets selective BCL-2 proteins and efficiently induces apoptosis via Bak/Bax if Mcl-1 is neutralized. Cancer Cell 2006; 10(5):389-99.
83. Del Gaizo Moore V, Schlis KD, Sallan SE et al. BCL-2 dependence and ABT-737 sensitivity in acute lymphoblastic leukemia. Blood 2008; 111(4):2300-9.

84. Deng J, Carlson N, Takeyama K et al. A.BH3 profiling identifies three distinct classes of apoptotic blocks to predict response to ABT-737 and conventional chemotherapeutic agents. Cancer Cell 2007; 12(2):171-85.

85. Konopleva M, Contractor R, Tsao T et al. Mechanisms of apoptosis sensitivity and resistance to the BH3 mimetic ABT-737 in acute myeloid leukemia. Cancer Cell 2006; 10(5):375-88.

86. Ishitsuka K, Yotsumoto F, Katsuya H et al. A promising therapeutic implication of a novel BCL-2 family inhibitor ABT-737 for adult T-cell leukemia/lymphoma. Blood (ASH Annual Meeting Abstracts) 2008; 112:1584.

87. Jayanthan A, Howard SC, Trippett T et al. Targeting the BCL-2 family of proteins in hodgkin lymphoma: in vitro cytotoxicity, target modulation and drug combination studies of the BH3 mimetic ABT-737. Blood (ASH Annual Meeting Abstracts) 2008; 112:3626.

88. Chauhan D, Velankar M, Brahmandam M et al. A novel BCL-2/Bcl-X(L)/BCL-2 inhibitor ABT-737 as therapy in multiple myeloma. Oncogene 2007; 26(16):2374-80.

89. Kline MP, Rajkumar SV, Timm MM et al. ABT-737, an inhibitor of BCL-2 family proteins, is a potent inducer of apoptosis in multiple myeloma cells. Leukemia 2007; 21(7):1549-60.

90. Paoluzzi L, Gonen M, Bhagat G et al. The BH3-only mimetic ABT-737 synergizes the antineoplastic activity of proteasome inhibitors in lymphoid malignancies. Blood 2008; 112(7):2906-16.

91. Kojima K, Konopleva M, Samudio IJ et al. Concomitant inhibition of MDM2 and BCL-2 protein function synergistically induce mitochondrial apoptosis in AML. Cell Cycle 2006; 5(23):2778-86.

92. Kang MH, Kang YH, Szymanska B et al. Activity of vincristine, L-ASP and dexamethasone against acute lymphoblastic leukemia is enhanced by the BH3-mimetic ABT-737 in vitro and in vivo. Blood 2007; 110(6):2057-66.

93. Kuroda J, Puthalakath H, Cragg MS et al. Bim and Bad mediate imatinib-induced killing of Bcr/Abl+ leukemic cells and resistance due to their loss is overcome by a BH3 mimetic. Proc Natl Acad Sci USA 2006; 103(40):14907-12.

94. Chen S, Dai Y, Harada H et al. Mcl-1 down-regulation potentiates ABT-737 lethality by cooperatively inducing Bak activation and Bax translocation. Cancer Res 2007; 67(2):782-91.

95. Ricciardi MR, Milella M, Libotte F et al. Synergistic induction of apoptosis in multiple myeloma cells by simultaneous inhibition of the Raf/MEK/ERK and BCL-2 pathways. Blood (ASH Annual Meeting Abstracts) 2008; 112:5161.

96. Trudel S, Stewart AK, Li Z et al. The BCL-2 family protein inhibitor, ABT-737, has substantial antimyeloma activity and shows synergistic effect with dexamethasone and melphalan. Clin Cancer Res 2007; 13(2 Pt 1):621-9.

97. Tahir SK, Yang X, Anderson MG et al. Influence of BCL-2 family members on the cellular response of small-cell lung cancer cell lines to ABT-737. Cancer Res 2007; 67(3):1176-83.

98. Lin X, Morgan-Lappe S, Huang X et al. 'Seed' analysis of off-target siRNAs reveals an essential role of Mcl-1 in resistance to the small-molecule BCL-2/BCL-XL inhibitor ABT-737. Oncogene 2007; 26(27):3972-9.

99. Olejniczak ET, Van Sant C, Anderson MG et al. Integrative genomic analysis of small-cell lung carcinoma reveals correlates of sensitivity to bcl-2 antagonists and uncovers novel chromosomal gains. Mol Cancer Res 2007; 5(4):331-9.

100. Xia L, Wang R, Qian H et al. Arsenic trioxide plus ABT-737 is synergistic to induce apoptosis in AML cells by repression of Mcl-1 protein and inactivation of BCL-2 protein. Blood (ASH Annual Meeting Abstracts) 2008; 112:2653.

101. Zhang H, Nimmer PM, Tahir SK et al. BCL-2 family proteins are essential for platelet survival. Cell Death Differ 2007; 14(5):943-51.

102. Wilson WH, Tulpule A, Levine AM et al. A phase 1/2a study evaluating the safety, pharmacokinetics and efficacy of ABT-263 in subjects with refractory or relapsed lymphoid malignancies. Blood (ASH Annual Meeting Abstracts) 2007; 110:1371.

103. Klasa RJ, Gillum AM, Klem RE et al. Oblimersen BCL-2 antisense: facilitating apoptosis in anticancer treatment. Antisense Nucleic Acid Drug Dev 2002; 12:193-213.

104. Cotter FE, Waters J, Cunningham D. Human BCL-2 antisense therapy for lymphomas. Biochim Biophys Acta 1999; 1489(1):97-106. Review.

105. O'Connor OA, Smith EA, Toner LE et al. The combination of the proteasome inhibitor bortezomib and the bcl-2 antisense molecule oblimersen sensitizes human B-cell lymphomas to cyclophosphamide. Clin Cancer Res 2006; 12(9):2902-11.

106. Smith MR, Jin F, Joshi I. Enhanced efficacy of therapy with antisense BCL-2 oligonucleotides plus anti-CD20 monoclonal antibody in scid mouse/human lymphoma xenografts. Mol Cancer Ther 2004; 3(12):1693-9.

107. Pro B, Leber B, Smith M et al. Phase II multicenter study of oblimersen sodium, a BCL-2 antisense oligonucleotide, in combination with rituximab in patients with recurrent B-cell nonHodgkin lymphoma. Br J Haematol 2008; 143(3):355-60.

108. O'Brien S, Wu J, Novick S et al. 5-year follow-up of patients with relapsed/refractory CLL treated with standard chemotherapy with or without oblimersen in randomized phase III trial: prognostic factors and predictive factors for treatment effect. Blood (ASH Annual Meeting Abstracts) 2008; 112:4201.

109. Chuntharapai A, Dodge K, Grimmer K et al. Isotype-dependent inhibition of tumor growth in vivo by monoclonal antibodies to death receptor 4. J Immunol 2001; 166:4891-4898.

110. Ichikawa K, Liu W, Zhao L et al. Tumoricidal activity of a novel anti-human DR5 monoclonal antibody without hepatocyte cytotoxicity. Nature Med 2001; 7:954-960.

111. Takeda K, Yamaguchi N, Akiba H et al. Induction of tumor-specific T-cell immunity by anti-DR5 antibody therapy. J Exp Med 2004; 199:437-448.

112. Younes A, Vose JM, Zelenetz AD et al. Results of a phase 2 trial of HGS-ETR1 (agonistic human monoclonal antibody to TRAIL receptor 1) in subjects with relapsed/refractory nonHodgkin's lymphoma (NHL). Blood (ASH Annual Meeting Abstracts) 2005; 106:489.

113. Shankar S, Singh TR, Fandy TE et al. Interactive effects of histone deacetylase inhibitors and TRAIL on apoptosis in human leukemia cells: involvement of both death receptor and mitochondrial pathways. Int J Mol Med 2005; 16(6):1125-38.

114. Inoue S, MacFarlane M, Harper N et al. Histone deacetylase inhibitors potentiate TNF-related apoptosis-inducing ligand (TRAIL)-induced apoptosis in lymphoid malignancies. Cell Death Differ 2004; 11:S193-S206.

115. Mitsiades CS, Treon SP, Mitsiades N et al. TRAIL/Apo2L ligand selectively induces apoptosis and overcomes drug resistance in multiple myeloma: therapeutic applications. Blood 2001; 98:795-804.

116. Ashkenazi A, Pai RC, Fong S et al. Safety and antitumor activity of recombinant soluble Apo2 ligand. J Clin Invest 1999; 104:155-162.

117. Ashkenazi A. Targeting death and decoy receptors of the tumour-necrosis factor superfamily. Nature Rev Cancer 2002; 2:420-421.

118. Kelley SK, Ashkenazi A. Targeting death receptors in cancer with Apo2L/TRAIL. Curr Opin Pharmacol 2004; 4:333-339.

119. Yee L, Fanale M, Dimick K et al. A phase IB safety and pharmacokinetic (PK) study of recombinant human Apo2L/TRAIL in combination with rituximab in patients with low-grade nonHodgkin lymphoma. Journal of Clinical Oncology, 2007 ASCO Annual Meeting Proceedings Part I 2007; 25(18S):8078.

120. Shimizu S, Kanaseki T, Mizushima N et al. Role of BCL-2 family proteins in a non-apoptotic programmed cell death dependent on autophagy genes. Nature Cell Biol 2004; 6:1221-1228.

121. Pattingre S, Tassa A, Qu X et al. BCL-2 antiapoptotic proteins inhibit Beclin 1-dependent autophagy. Cell 2005; 122(6):927-39.

122. Levine B, Kroemer G. Autophagy in the pathogenesis of disease. Cell 2008; 132:27-42.

123. Yoshimori T. Autophagy: paying Charon's toll. Cell 2007; 128:833-836.

124. Janumyan Y, Cui Q, Yan L et al. G0 function of BCL2 and BCL-xL requires BAX, BAK and p27 phosphorylation by Mirk, revealing a novel role of BAX and BAK in quiescence regulation. J Biol Chem 2008; 283(49):34108-20.

125. Cui Q, Valentin M, Janumyan Y et al. Bax(-/-) bak(-/-) cells exhibit p27 Thr198 phosphorylation and autophagy. Autophagy 2009; 5(2).

126. Yazbeck VY, Buglio D, Georgakis GV et al. Temsirolimus downregulates p21 without altering cyclin D1 expression and induces autophagy and synergizes with vorinostat in mantle cell lymphoma. Exp Hematol 2008; 36(4):443-50.

CHAPTER 6

The Interplay between BCL-2 Family Proteins and Mitochondrial Morphology in the Regulation of Apoptosis

Maria Eugenia Soriano and Luca Scorrano*

Abstract

Apoptosis is a highly regulated process where key players such as BCL-2 family members control the recruitment of the mitochondrial subroutine. This culminates in the release of cytochrome *c* from the organelle in the cytoplasm, where it is required for the activation of effector caspases. The complete release of cytochrome *c* is the result of the combined action of proapoptotic BCL-2 family members and of changes in the complex morphology and ultrastructure of the organelle, controlled by the balance between fusion and fission processes. Here we discuss recent findings pointing to a role for changes in mitochondrial morphology during apoptosis and how these might be regulated by members of the BCL-2 family.

Introduction

The BCL-2 family proteins are critical regulators of apoptosis, acting as central players in the mitochondrial pathway of cell death. The BCL-2 family is composed by anti- and proapoptotic proteins structurally characterized by the presence of four distinctive BCL-2 homology (BH) domains, crucial to confer anti- or proapoptotic function to the different family members. Antiapoptotic proteins share homology in all BH1 to 4 segments, while proapototic members lack stringent sequence conservation of the first BH4 domain. Proapoptotic proteins can be further subdivided into "multidomain" and "BH3 only." The multidomain proapoptotic proteins display sequence conservation in BH1-3 domains, while the BH3 only conserve only the BH3 region. Most of the BCL-2 family proteins contain a C-terminal membrane anchoring sequence that localizes them to intracellular membranes (mainly to mitochondria) where they execute their function.

An Historical Perspective on the Proteins of the BCL-2 Family

The term "apoptosis" was used first in 1972 by John Kerr, Andrew Wyllie and Alistair Currie to describe a little known and curious form of cell death—a programmed cell suicide—that was distinctly different from the long-recognized process of cell death known as necrosis. After a long period of silence, during the last 15 years, intense research confirmed that apoptosis plays a central role within developing organisms by shaping the neural and immune systems and moulding tissue specificity. Moreover, apoptosis helps to establish a natural balance between cell death and cell

*Corresponding Author: Luca Scorrano—Department of Cell Physiology and Metabolism, University of Geneva Medical School, 1 Rue M. Servet, 1211 Genève, Switzerland. Email: luca.scorrano@unige.ch

BCL-2 Protein Family: Essential Regulators of Cell Death, edited by Claudio Hetz.
©2010 Landes Bioscience and Springer Science+Business Media.

renewal in mature tissues by destroying excess, damaged, or abnormal cells. It appears intuitive that apoptosis could be also involved in cancer. However, the link between apoptosis and cancer was not established until David Vaux and colleagues demonstrated in 1988 that the *bcl-2* gene specifically blocks death of B-cells in follicular lymphoma.[1] The BCL-2 protein is overexpressed in these tumour cells and the gene was identified at the chromosomal breakpoint of t(14;18) bearing B-cell lymphomas.[2] In 1990, Stanley Korsmeyer, David Hockenbery and colleagues characterized the normal *bcl-2* as a suicide "brake" gene, producing a protein that blocks apoptosis.[3] *BCL-2* was the first antideath gene discovered, a milestone that effectively launched a new era in cell death research. Up to that moment, most of the knowledge concerning oncogenic events were focused in the mechanism of increased growth and proliferation. Later studies showed that BCL-2 and other antiapoptotic members of the family are overexpressed in several types of cancer and demonstrated that they mainly promoted cell survival but not proliferation, leading to the conclusion that impaired apoptosis is a critical step in tumour development.[4,5]

Apoptotic Pathways

Programmed cell death and cell proliferation are mandatory process during embryonic development and in the maintenance of homeostasis in normal tissues. A good example is the immune system where cell selection and differentiation in mammals are regulated by the convergence of survival and death signals to eliminate most of the cells lacking the right features. In general, defects in genes that control these pathways provoke the survival of cells normally destined to die, resulting in both cancer and autoimmune diseases, while defects causing excessive cell death may contribute to degenerative diseases and immunodeficiency. The same picture can be applied to other tissues and organs.

There are two main pathways of cell death, the extrinsic and the intrinsic pathway, that differ for the relative role of mitochondria.[6,7] The extrinsic pathway is mediated by the activation of the so called "death receptors" and it is important in the maintenance of tissue homeostasis, being triggered by the neighbouring environment. Receptors oligomerize after ligand interaction (i.e., between TNFα and TNF-RI or II) forming the so called death inducing signalling complex (DISC), which results in caspases-8 activation and subsequent activation of downstream effector caspases-3.[8-10] In some cell types like lymphocytes the level of activation of apical caspases-8 is not sufficient to generate enough caspases-3 activation and mitochondria come into place as key amplifiers of the death signal. Activated caspases-8 cleaves the proapoptotic protein Bid which translocates to mitochondria and contributes to cytochrome *c* mobilization and releases to activate new proapoptotic proteins.[11-13] On the other hand, the intrinsic pathway uses mitochondria as the main active players and starting point of the death signalling events triggered by DNA damage,[14-16] growth factor withdrawal[17] or hypoxia.[18] In any case, mitochondria play an essential role in cell death and is at the mitochondrial level that most of the BCL-2 family proteins execute their function, inducing or inhibiting the cell commitment to death as will be described later in detail.

BCL-2 Family

BCL-2 family proteins have been identified by homology screening for their BCL-2 homology domains from the first identified protein BCL-2 (Fig. 1). The presence of 1 up to 4 BH domains in these proteins is a common feature and most of them posses a C-terminal transmembrane domain that anchors them to intracellular membranes of mitochondria, ER and nucleus.[19] Ever since the discovery of BCL-2, the number of new members has dramatically increased, revealing different patterns of developmental expression, subcellular localization and responsiveness to specific death stimuli. BCL-2 family proteins have been classified according to their BH composition and function in "antiapoptotic" and "proapoptotic" members, the latter ones subdivided in "multidomain" and "BH3 only" proteins. A widely accepted model postulates that the stoichiometry between pro and anti-apoptotic BCL-2 proteins commits to cell survival or death.[20,21]

In general, most of the antiapoptotic proteins in mammals like BCL-2, BCL-XL, Bcl-W contain all BH 1-4 domains; proapoptotic multidomain proteins like Bax, Bak and Bok display BH1 to 3

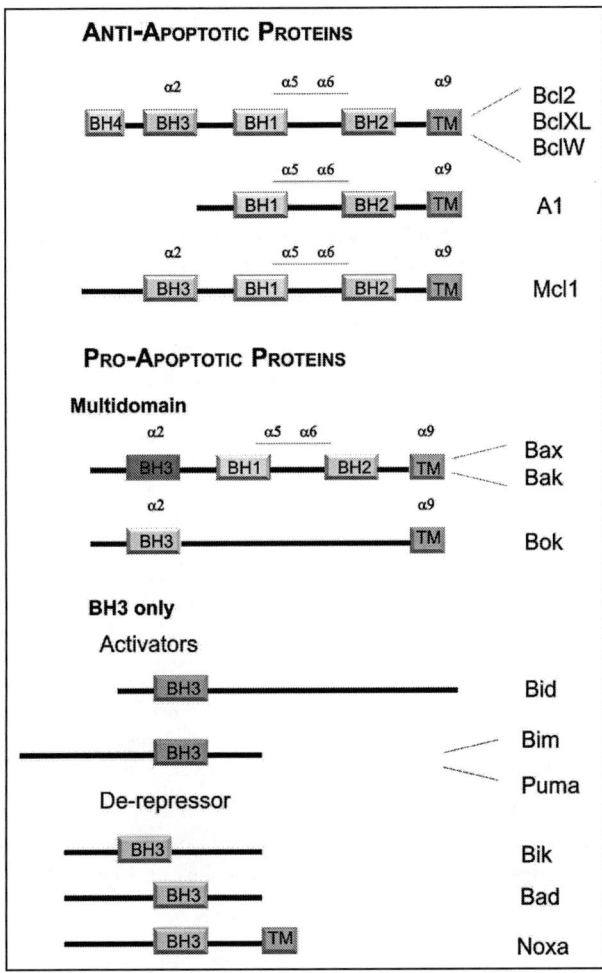

Figure 1. BCL-2 family proteins. A scheme illustrating the overall structural arrangement of anti and pro-apoptotic BCL-2 family members. The Bcl2-homology (BH) domains are highlighted.

domains, while proapoptotic BH3 only proteins show only the BH3 domain. From a structural point of view, these proteins contain several alpha-helix some of which are important for membrane insertion (α-9, in the TM domain), interaction with BH3 domain (α-2) and pore formation (α-5 and α-6, between BH1 and BH2 domains).[22-25] The structural homologies of some BCL-2 proteins to bacterial toxins, especially in α-5 and α-6 regions, led to hypothesize that they could form ion- or protein-conducting channels to explain their mechanism of action.[26] Several papers have been published demonstrating the ability of BCL-XL, BCL-2 and Bax to form channels upon self-oligomerization. The channels formed can be described as multiconductance, pH-sensitive, voltage-gated channels with poor ion selectivity. While BCL-XL and BCL-2 are cation selective and show this property only in vitro at acid pH, Bax forms ion channels anion selective in liposomes and phospholipids bilayers at physiological pH suggesting such activity in cells.[23,24] The differences in the electrophysiological properties could be due to the different aminoacidic composition of the α-5 and α-6 helices.[27,28] As we will described more in detail, the channel formation is not required for the apoptotic activity of these proteins in mammals.

BH Domains and Molecular Interaction

The molecular mechanism of action of the BCL-2 family proteins depends mainly on the BH domains. Interaction between pro- and antiapoptotic proteins depends on the BH domains. In the case of antiapoptotic proteins, they structurally display an hydrophobic groove formed by BH1,2 and 3 where BH4 may interact from the back stabilizing the structure and hiding some hydrophobic residues. These domains do not have an enzymatic activity but mediate the interaction of the BCL-2 antiapoptotic members with other BCL-2 proteins, being essentials for the survival function. The hydrophobic groove is the site of interaction with BH3 domains from other members of the family, resulting in homo- or heterodimerization. In fact, mutations within the BH domains result in a loss of the antiapoptotic functions in the antiapoptotic proteins, while in proapoptotic proteins mutations in the BH3 domain nullified the pro-death activity.[29,30] For example, BCL-XL complex with the proapoptotic protein Bax inserting the BH3 domain in the hydrophobic pocket of BCL-XL through hydrophobic and charge interactions with well known aminoacids in BH1, 2 and 3. Mutations in aminoacids present in the hydrophobic groove of BCL-XL suppress its antiapoptotic activity probably because Bax or Bak are not able to interact and remain free to be activated instead that sequester by BCL-XL. This model is well accepted at least in vitro, because structural analysis of BH3 domains from other proteins confirmed the presence of hydrophobic and charged amino acids placed properly in most of the BH3 containing proteins.[31-34] The above described interaction result theoretically in formation of dimmers or oligomers of antiapoptotic proteins with multidomain or BH3 only proapoptotic proteins. However, proapoptotic proteins as Bax and Bak, with available BH3 domains, not only interact with BCL-2 antiapoptotic proteins, but also form oligomers in the mitochondrial membrane, while proapoptotic BH3-only proteins somehow interact with both pro- and antiapoptotic proteins. The balance in the expression of pro- and antiapoptotic proteins determines their homo- or heterodimerization and therefore the fate of the cell.

The presence of the BH4 domain only in antiapoptotic members led to propose that this domain plays a critical role in the antiapoptotic function. However, it remains unclear how proapoptotic members as Mcl-1 and A1 lacking the BH4, execute their function.

BCL-2 Antiapoptotic Proteins

The antiapoptotic BCL-2 members are characterized by the presence of all the conserved BH1 to 4 domains. Most localize on the cell membranes, mainly in mitochondria, where they inhibit cytochrome *c* release. The antiapoptotic BCL-2 proteins known in mammals are BCL-2, BCL-XL, Bcl-W, Mcl-1, A1, DIVA and NR13. The analysis of the three-dimensional structures of the BCL-2 antiapoptotic proteins described up to now showed a high degree of similarity. They display two central hydrophobic α-helices surrounded by six or seven amphipathic α-helices. In the surface of BCL-2, BCL-XL and Mcl-1 there is a hydrophobic task that is the binding site for the BH3 region of pro-apoptotic BCL-2 family members, which in this way annul the antiapoptotic effect.

BCL-2 and BCL-XL

The over expression of BCL-2 and BCL-XL has been associated with several types of tumors, including chronic lymphocytic leukaemia, multiple myeloma, melanoma, prostate, ovarian, cervical, bladder, gastric, pancreatic, breast and colorectal cancers. In most circumstances, the overexpression of these proteins correlates with poor survival and disease progression. During development, the different patterns of expression of these proteins contribute to the survival of specific cells. This occurs during B- and T-cell development. The earliest progenitors express BCL-2, but it is downregulated by the critical stage of lymphocyte development when cells with non functional and autoreactive antigen receptors are eliminated by apoptosis. If BCL-2 expression is altered, defectives lymphocytes will not be eliminated, promoting an abnormal development.[35-39] In fact, *bcl-2* deficient mice demonstrate fulminant lymphoid apoptosis among other alterations (polycystic kidneys and hypopigmented hair, motoneuron degeneration and disordered growth of intestinal villi and long bones) and BCL-2 over expression has been shown to promote chronic lymphocytic leukaemia.[40-42]

More dramatic effects are found in *bcl-XL*-null mice, which display massive apoptotic death of immature neurons throughout the developing nervous system and die around embryonic day 13.5 from hematopoietic cell failure. BCL-XL is expressed at very low levels in neuronal precursor cells but is upregulated upon their migration from the ventricular zone and differentiation and remains high in mature, postmitotic CNS neurons. Several studies have revealed the important role of BCL-XL in promoting neuron survival in developing and adult CNS.

Bcl-W

Bcl-W is a pro-survival protein which when overexpressed renders myeloid and lymphoid cell lines refractory to apoptosis induced by cytokine deprivation or irradiation, but is relatively ineffective against apoptosis induced by engagement of the CD95 (Fas) death receptor.[43] Transcripts of *bcl-w* are present at moderate levels in brain, colon and salivary gland and at low levels in testis, liver, heart, stomach, skeletal muscle and placenta, as well as in most myeloid cell lines but few lymphoid lines.[44-49] Mice that lack *bcl-W* are viable, healthy and normal in appearance. Most tissues exhibit typical histology and hematopoiesis is unaffected, presumably due to redundant function with other pro-survival family members. Interestingly, the males are infertile due to disorganization of the seminiferous tubules, containing numerous apoptotic cells. Both Sertoli and germ cells of all types are reduced in number, with a gradient of depletion towards maturation.[50] This supports a role for this protein in the specific development of another mitochondria-rich cell.

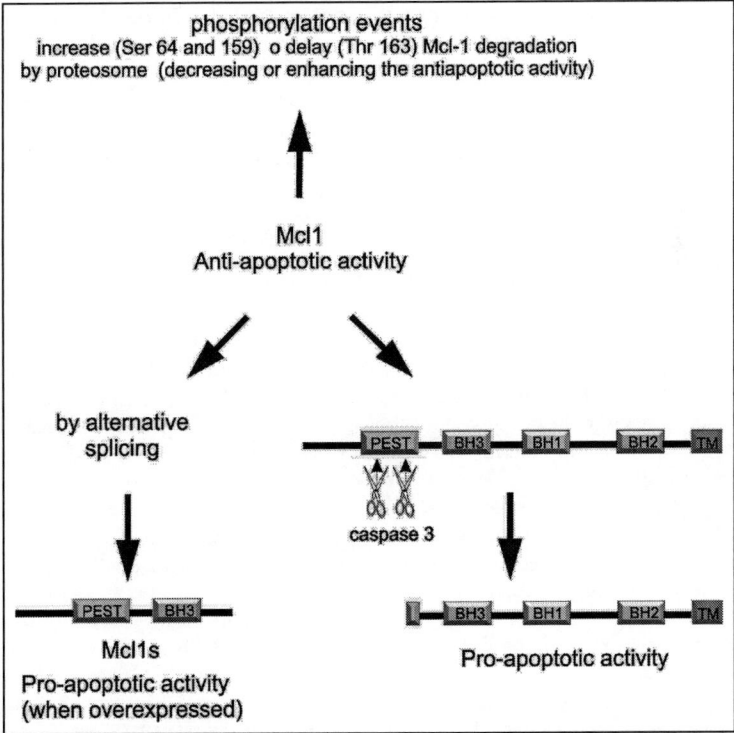

Figure 2. Regulation of Mcl-1 function and degradation. Mcl-1 antiapoptotic activity may become proapototic (i) by alternative splicing and elimination of the BH1-2 domains and transmembrane region; (ii) or after caspase cleavage in the PEST region. In addition, antiapoptotic activity is regulated by the proteasome and Mcl-1 degradation.

Mcl-1

The Mcl-1 (myeloid cell leukemia-1) protein is a member of the BCL-2 family lacking the "antiapoptotic" BH4 domain, but that still conserves potent antiapoptotic activity (Fig. 2). Mcl-1 also contains a C-terminal transmembrane domain that serves to localize it to various intracellular membranes, mainly to the mitochondrial outer membrane; and in the N-terminal the protein display two PEST regions (Proline (P), glutamate (E), serine (S) and threonine (T) rich regions).[51] Such regions are typical of proteins with a short half-life that are degradated via proteasome. In fact, Mcl-1 is a highly regulated protein induced by a wide range of survival and differentiation signals such as cytokines and growht factors. Mcl-1 is also rapidly downregulated during apoptosis.[52] The survival function of Mcl-1 can be also altered by effector caspases cleavage after two aspartic acid residues. The C-terminal fragment resulting from cleavage is a potent cell death promoting protein, so caspase cleavage of Mcl-1 simultaneously deprives cells of a survival protein and generates a proapoptotic molecule. In addition alternative splicing of Mcl-1 gives rise to a second protein isoform, Mcl-1s, which lacks the BH1, 2 and transmembrane domain but conserves PEST and BH3 regions. Overexpression of this isoform also promotes cell death.

Mcl-1 seems to inhibit cell death by inhibiting cytochrome *c* release. In vitro studies have shown that Mcl-1 interacts with high affinity to the BH3 only proteins Bid, Bim and Puma, but also selectively with Noxa. Further studies suggested that Mcl-1 may sequester Bak in the inactive form preventing its oligomerization.[53-55]

Mcl-1 is expressed in a wide variety of cell types, in the adult and during embryonic development with different patterns tissue and differentiation specific. In the epithelium Mcl-1 is expressed highly in the more differentiated layers of ephitelia. In the lymphoid system is essential in the development and maintenance of B- and T-lymphocytes.[53,56,57]

Mcl-1 null phenotype is a combination of an embryonic developmental delay and an implantation lethal defect of mouse embryos.[57] Conditional knockout models further indicate that Mcl-1 is required for the development and maintenance of B- and T-lymphocytes and to ensure the homeostasis of early hematopoietic progenitors.[53,58] Leukemic cells display an elevated expression of Mcl-1, acquiring resistance to chemoterapeutic agents.[59] In fact, Mcl-1 transgenic mice show high incidence of B-cell lymphoma.[60]

Bfl-1/A1

Bfl1 is one of the smallest members of the Bcl2 family with 175 aa. Human and mouse protein share a BH1-, BH2- and somewhat less conserved BH3 domain, but the limited homology to a BH4 domain found in human Bfl1 is not present in the mouse A1. Protection by Bfl-1 impinges on intact BH1- and BH2 domains and the presence of a TM domain determines its localization on the OMM but it seems not to be necessary for the antiapoptotic activity. While there are four *a1* genes in the mouse genome, the unique human *bfl-1* gene encodes a full-length protein that localizes at mitochondria as well as an alternatively spliced variant (Bfl-1$_S$) lacking 12 C-terminal amino acids that localizes to the nucleus but is nevertheless protective against apoptosis.[61] Bfl-1 transcripts are constitutively expressed in bone marrow, lymphoid organs and peripheral blood leukocytes and are induced in response to activation of NF-KB transcription factors in many cell types. These findings suggest an important role for Bfl-1 in the survival of cells in the immune system.

As Mcl-1, Bfl-1 is highly regulated and may act as antiapoptotic molecule suppressing cell death induced by staurosporine and cytokine withdrawal.[62,63] Bfl-1 becomes proapoptotic after proteolytic cleavage by caspases, calpains or molecules with a similar activity triggered by stimulation with TNFα and cycloheximide (CHX).[64] It has been proposed that an inhibition of Bfl-1 proteasome degradation also contributes to the proapoptotic activity. In fact, deletion of the C-terminal domain which is necessary for ubiquitination and proteasome mediated turnover, counteract the proapoptotic effects associated with TNFα/CHX.

The mechanism of action of Bfl-1it is unclear, Bfl-1 has been shown to bind Bax in yeast two-hybrid system but is not able to form homodimers nor to interact with human BCL-2 or

BCL-XL. Bax requires the BH3 domain to interact with Bfl-1 and the latter requires the same domains that form the majority of the hydrophobic pocket in BCL-XL.[65] Alternatively, Bfl-1 could act by inhibiting the collaboration between the BH3 -only protein Bid and its proapoptotic partners Bax or Bak in the induction of cytochrome *c* release.[66,67] Bfl-1 does so by binding to the BH3 domain of full-length Bid. It does not interfere with proteolytic activation of Bid, nor with its mitochondrial insertion, but remains selectively complexed to tBid in the mitochondrial membrane where it prevents the activity of a proapoptotic complex.[68]

BCL-2 Proapoptotic Proteins

As previously mentioned, the BCL-2 proapototic members can be subdivided into multidomain and BH3-only proteins. Multidomain proteins comprise members Bax, Bak, Bok, Boo/DIVA and are characterized by sequence conservation in BH1 to 3 domain, while BH 3 only proteins display only the BH3 domain.

Bax and Bak

Bax and Bak are the "executers" or "effectors" members of the BCL-2 family since they are the direct active players in cytochrome *c* release which leads to caspases activation and cell demise. They differ in their intracellular localization before a death stimulus, Bak being an integral protein of the mitochondrial outer membrane, while Bax in viable cells resides as a monomer in the cytosol or attached to intracellular membranes.[69,70] In response to apoptotic stimuli, Bax or Bak induce OMM permeabilization leading to release of proteins normally retained in the mitochondrial intermembrane space as cytochrome *c*.[71,72] A widely accepted model propose that during apoptosis Bax and Bak are activated and undergo a conformational change that lead to membrane localization (in the case of Bax), oligomerization and formation of pores trough which cytochrome *c* is released.[73-75] There are several hypothesis to explain Bax and Bak activation: (i) they are directly activated by BH3 only proteins; (ii) a de-repressor BH3 only protein competes for binding to a BCL-2 antiapoptotic member that had a BH3 only activator protein sequestered, resulting in the release to activate Bax and Bak; (iii) the BH3 only non activator proteins sequester BCL-2 antiapoptotic proteins freeing the BH3 only activator proteins to activate Bax and Bak (Fig. 3).

From a structural point of view, in solution the BH1-3 domains of Bax form an hydrophobic pocket similar to that of BCL-2 and BCL-XL, with the C-terminal tail folded back into the hydrophobic pocket of the molecule.[69] Release of the C-terminal tail provokes mitochondrial targeting and a conformational change that expose the BH3 domain allowing the interaction with other family proteins (activated conformation). It is unclear if BH3 only proteins trigger the conformational change in Bax by competition by the binding pocket or some other factors or events induce it. The conformational change results in insertion in the OMM via the pore forming α-5 and α-6 helices and probably Bax oligomerization to form pores[76,77] (Fig. 4). Even if Bax oligomers have been found in vitro and after crosslinking, clear evidence of the existence of Bax oligomers in vivo is lacking. Interestingly, immunoelectromicroscopy experiments unveiled large clusters of Bax and Bak around mitochondria both in Bax overexpressing and apoptotically stressed cells.[78] Interestingly, these Bax/Bak clusters are proximal to sites where mitochondria appear to fragment, suggesting a possible interplay between members of the BCL-2 family and the regulation of mitochondrial morphology. We will discuss this possibility in detail.

BH3 Only Proteins

These group of proteins are characterized by the homology with the other BCL-2 family members only in the short BH3 domain. They are present as inactive forms in healthy cells and are activated upon an apoptotic stimulus. Among the events that result in BH3 only proteins activation we find transcriptional induction, phosphorylation and proteolysis. For example, Noxa and Puma expression is induced by the transcription factor p53, which is activated in apoptosis induced by DNA damage,[79,80] Bim is induced by the transcription factor FOXO3A (class O forkhead box transcription factor-3A) after growth factor deprivation, or by CEBPβ

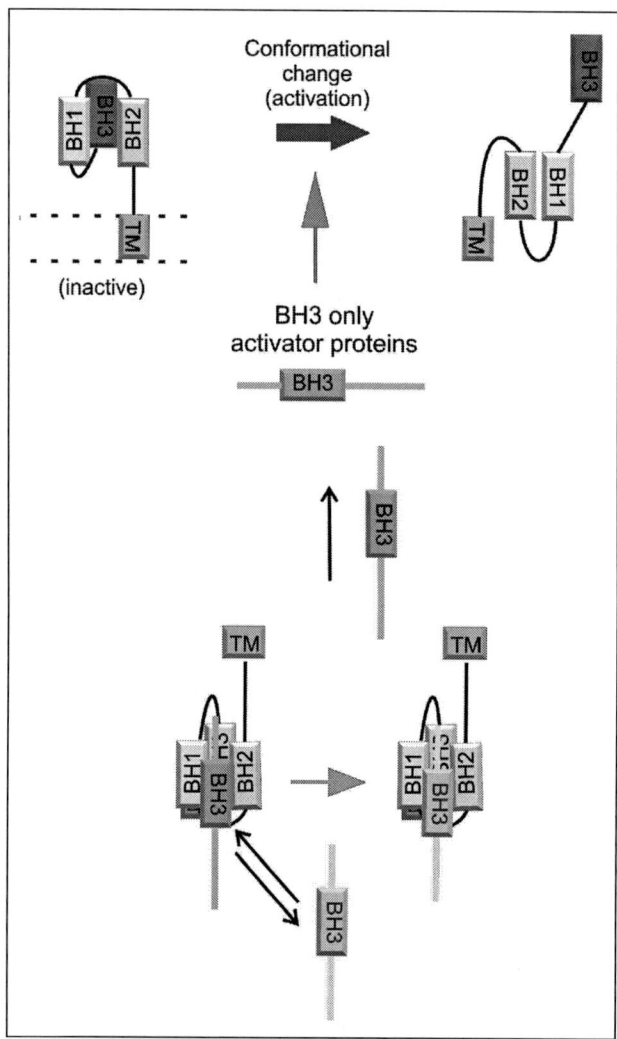

Figure 3. Models of Bax and Bak activation and oligomerization. After an apoptotic stimulus Bax or Bak suffer a conformational change (activation event) that leads to exposure of the BH3 domain. Self-oligomerization occurs probably through the interaction of BH3 domain of Bax and Bak, allowing the formation of a channel and cytochrome c release.

or CHOP during reticulum endoplasmic stress. Bim can be also activated by release from dynein motor complex after growth factor deprivation or UV irradiation, remaining free to sequester BCL-2 antiapoptotic proteins. The expression of Bid has been shown to be regulated by p53 in thymocytes.[81] Hrk/DP5 has been suggested to be induced in damaged or stressed neurons by the JNK pathway. Bad is normally phosphorylated at several serine residues that allows its sequestration in the cytoplasm by binding to the 14-3-3 scaffold protein. When dephosphorylated, Bad is released from 14-3-3 to interact with other BCL-2 family members to carry out its proapoptotic function.[82,83] In the case of Bik, its phosphorylation increases the proapoptotic activity by an unknown mechanism. Finally, Bid which is normally in the cytosol

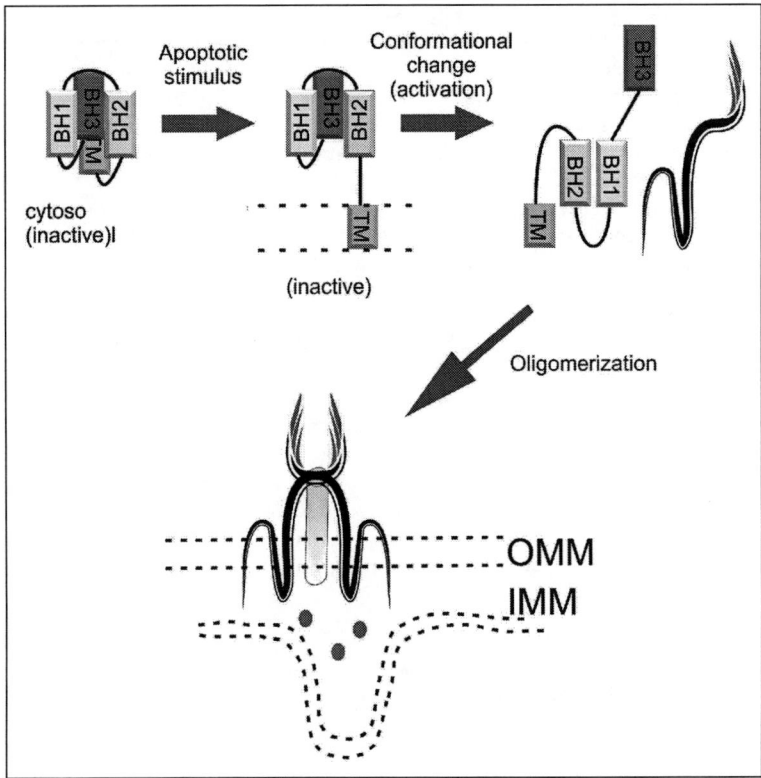

Figure 4. Activation of Bax or Bak by BH3 only proteins. After an apoptotic stimulus, Bax and Bak can be activated: (i) directly by a BH3 only activator protein (green BH3 domain); (ii) a de-repressor BH3 only protein (yellow BH3 domain) competes for binding to a BCL-2 antiapoptotic member that had a BH3 only activated protein sequestered, which would be able to activate Bax and Bak; (iii) the BH3 only non activator proteins bind to antiapoptotic protein remaining free the BH3 only activator proteins to activate Bax and Bak. A color version of this image is available at www.landesbioscience.com/curie.

is activated by cleavage of caspase 8 in the death pathway mediated by membrane receptors. On the contrary, phosphorylated Bid is resistant to cleavage by caspases in in vitro assays.[84] Some data have suggested Bid cleavage by other caspases than caspase-8 and a possible role in other types of cell death not mediated by receptor.[12,85] The truncated form of Bid (tBid) translocates to mitochondria to carry out its death function.

The BH3 only proteins can be further subdivided functionally into two distinct groups: one comprises the ones that sensitize cells to apoptosis by binding to BCL-2 antiapoptotic proteins (Noxa, Bik and Bad) and a second one that includes the direct activators of Bax and Bak (Bid, Bim and Puma).[86] In mammals there are at least 7 BH3 only proteins and each member seem to be critical in the tissue homeostasis and development. A good example is given by the role of Bim and Bid in T-cell differentiation:[81] *Bim* knock out animals develop lymphoproliferative diseases as leukemia and are resistant to cell death induced by cytokines and growth factor withdrawal. As we will see later, similarly to the other members of BCL-2 family BH3-only proteins also participate in the regulation of mitochondrial morphology.

Mitochondrial Shape and Apoptosis

Classically, mitochondria were regarded as essential organelles for the propagation of the apoptotic cascade, via the release in the cytoplasm of cytochrome *c* and other proapoptotic factors that are required for the activation of effector caspases. However, these profound changes were normally disjoint from morphological alterations of the organelle. A growing body of recent evidence suggests that mitochondrial morphology and ultrastructure is conversely altered during apoptosis, pointing to an interplay between critical death regulators of the BCL-2 family and the shape of mitochondria. We will now overview the main proteins regulating mitochondrial morphology, the morphological changes that occur in mitochondria during apoptosis and the role played by BCL-2 family members in the modulation of mitochondrial shape in healthy and dying cells.

Mitochondrial Morphology: An Equilibrium between Fusion and Fission

The functional versatility of mitochondria is paralleled by their morphological complexity. In certain cell types mitochondria are organized in networks of interconnected organelles.[87] Ultrastructurally, the inner membrane (IM) can be further subdivided in an inner boundary membrane and in the cristae compartment, bag-like folds of the IM connected to the intermembrane space (IMS) via narrow tubular junctions.[88] The ultrastructure and the reticular organization of the organelle are determined by mitochondria-shaping proteins that impinge on the equilibrium between fusion and fission processes. Most of the regulators of mitochondrial morphology were originally found and characterized in yeast, while their mammalian counterparts have been identified only in the last 8 years. We will now overview the core components of the machinery that regulates mammalian mitochondrial shape (Fig. 5).

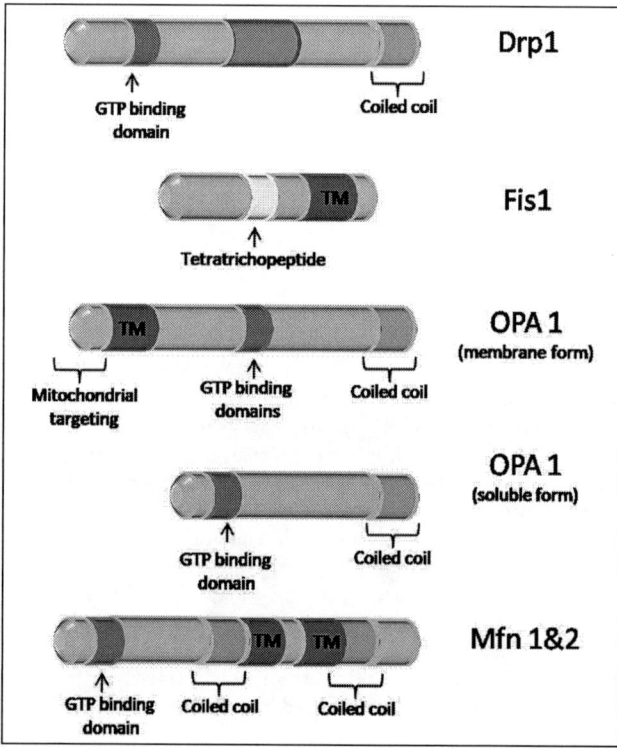

Figure 5. A cartoon depicting the structural arrangement of known core mitochondria-shaping proteins. The different domains are highlighted.

In mammalian cells, **mitochondrial division** is regulated by Drp1 and Fis1.[89-91] Drp1 is a cytosolic dynamin-related protein whose inhibition or downregulation result in a highly interconnected mitochondrial network. It shuttles to mitochondria in response to a variety of stimuli that ultimately impinge on calcineurin that dephosphorylates Drp1 to drive its translocation to the organelle.[92] Drp1 is recruited to mitochondria and constriction of the membranes takes place by direct or indirect interaction with Fis1[93] a 16 kDa integral protein of the outer mitochondrial membrane whose downregulation results in the elongation of the network.[91] Mitofusins (Mfn) 1 and 2 control **mitochondrial fusion** in mammals.[94,95] Deletion of either Mfn impairs mouse embryonic development at different stages, but conditional ablation of Mfn2 seems to be more severe than that of Mfn1.[96,97] Opa1 exists in eight different splice variants, is posttranscriptionally cleaved to generate at least 5 major species that can be retrieved associated to the inner membrane or soluble in the intermembrane space.[98] Opa1 promotes fusion in a process requiring Mfn1.[99]

Changes in mitochondrial shape are likely under the control of complex signaling cascades, as suggested by the recent finding of at least two different layer of regulation (Fig. 6). First, kinases and phosphatases impinge on mitochondrial morphology. Drp1 can be phosphorylated at two different sites by protein kinase A (PKA) and by cyclin dependent kinase 1 with opposing effect on morphology: the former promotes fusion,[100,101] the latter fission.[102] Action of PKA is counterbalanced by the phosphatase calcineurin to drive translocation of Drp1 to mitochondria.[92] Once translocated to the organelle, Drp1 can be stabilized by sumoylation[103] in a process that requires the proapoptotic BCL-2 family members Bax and Bak.[104] Last, but not least, mitochondrial movement is tightly controlled and coordinated with changes in the fusion/fission equilibrium. In mammals, mitochondria move along microtubules, thanks to two set of proteins that mediate antero- (kinesin) and retrograde (dynein) transport of the organelle. Ca^{2+} signals, the mitochondrial proteins Miro1 and 2 and the mitochondria-shaping protein Drp1, through its complex with dynamitin, appear to integrate morphological changes with the regulation of the transport of the organelle.[105] This suggests that forces

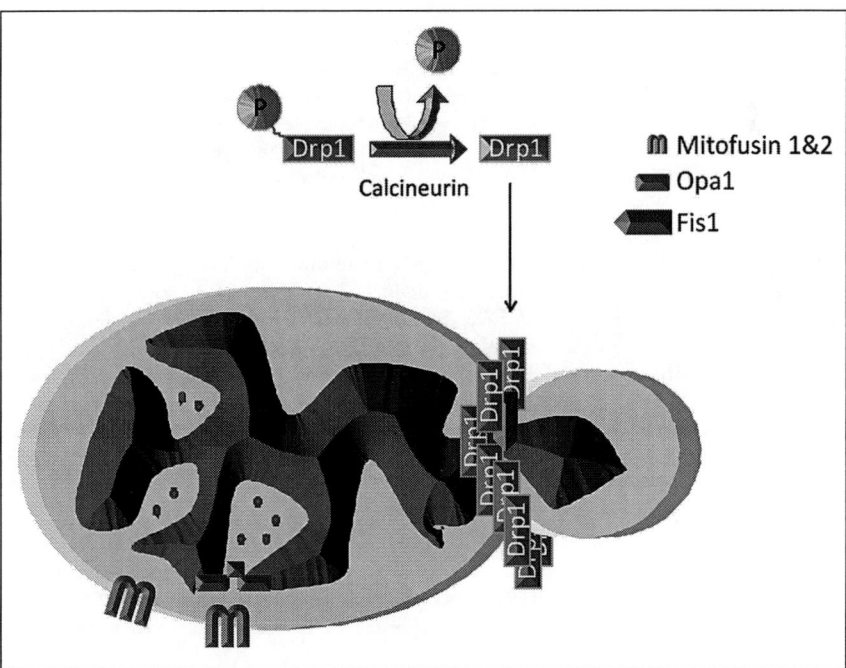

Figure 6. A cartoon of the regulation of mitochondrial fission by Drp1 and of mitochondrial cristae biogenesis and fusion by Opa1/Mfns.

exerted by molecular motors can be transmitted to mitochondrial membrane and participate in the regulation of the shape of the organelle.

Mitochondrial Fragmentation during Apoptosis: The Role of BCL-2 Family Members

The concept that mitochondria remain untouched during apoptosis changed since the discovery that reversible fragmentation occurs in neurons primed to die by growth factor deprivation.[106] This was then extended to apoptosis induced by several types of intrinsic stimuli, including Bax and linked to the progression of the apoptotic cascade. Inhibition of fragmentation by a dominant negative Drp1 mutant delayed release of cytochrome *c* and apoptosis.[107] Notably, not only mitochondrial fission is stimulated during apoptosis, but fusion is inhibited;[108] fission crucially contributes to death in neurons[109-113] and in cardiomyocytes[114,115] and to generation of reactive oxygen species;[116] interfering with the fusion-fission machinery affects selectively the release of cytochrome *c*, while that of other proapoptotic factors such as Smac is untouched.[117,118]

Despite our increased knowledge on the regulation of mitochondrial fragmentation during cell death, the debate is still open as to whether and how mitochondrial fission participates in the decisional phase of mitochondrial apoptosis, i.e., in the release of cytochrome *c*. The first and foremost problem is that we currently lack a mechanism to explain how enhanced fission might trigger or increase the release of proapoptotic factors from mitochondria. It has been postulated that sites of fission are preferential platforms for Bax/Bak oligomerization, but only indirect evidence of accumulation of Mfn2, Drp1 and Bax at discrete foci on mitochondrial surface exists.[78] The concept that proapoptotic members of the BCL-2 family can regulate the shape of mitochondria at the steady state is further reinforced by the finding that ablation of both Bax and Bak results in shorter mitochondrial tubuli, via an impaired regulation of Mfn2 dependent fusion karbowski nature. Whether this function is directly mediated by the physical interaction between Mfn2 and Bax/Bak, or via protein adaptors or mediators, remains to be assessed.

Chemical inhibition of Drp1 inhibits release of cytochrome *c* and death. However, this inhibitor is active also in an in vitro system of cytochrome *c* release, where levels of Drp1 associated with purified organelles are expected to be extremely low, raising the question of whether its action relies on the blockage of other mitochondrial pathways.[119] In addition, mitochondrial fission is the only required mitochondrial change that participates in the amplification of developmental apoptosis in lower eukaryotes where release of mitochondrial cofactors such as cytochrome *c* is not required.[120,121] Fission could be a part of an ancestral subroutine of mitochondrial elimination during death, that in mammals evolved to provide an amplificatory mechanism for the release of mitochondrial proteins.

On the contrary, it has been reported that changes in mitochondrial connectivity induced by multidomain proapoptotics like Bax can be dissociated from the release of cytochrome *c*: in particular, while the antiapoptotic BCL-XL blocks the release induced by Bax, it is apparently unable to prevent the changes in mitochondrial morphology induced by the enforced expression of the latter.[122] This report challenges several studies showing that high levels of pro-fusion proteins interfere with death induced by intrinsic stimuli as well as by conditions mimicking ischemia-reperfusion or other forms of tissue damage.[100,123-125] To conclude that BCL-XL does not antagonize fragmentation by Bax, one would expect that BCL-XL per se has no major effect on the morphology of the organelle. However, extensive literature supports a role for BCL-XL and other members of the BCL-2 family in the regulation of mitochondrial shape.[122,126-128] On the other hand, it appears that the antiapoptotic member Mcl-1 has no effect on mitochondrial dynamics.[122] Along this line, fragmentation induced by activating BH3 only such as Bid and Bim is almost completely blocked by BCL-2 and Mcl-1,[122] further corroborating a different interpretation of these results: fragmentation and cytochrome *c* release are tightly associated and can be disjoint only under special circumstances, such as the combination of anti- and pro-apoptotics with primary effects on mitochondrial morphology, or whose reciprocal binding is influenced by the tone of the endogenous other members of the BCL-2 family. In conclusion, we are far from understanding whether and how mitochondrial fission is required for the amplification of apoptosis in mammals. The field requires the establishment of conditional mouse models of ablation of pro-fission proteins and a genetic analysis of the occurrence of mitochondrial fragmentation during death induced in cells where cytochrome *c* release and activation of caspases is blocked. In addition, the interplay between BCL-2 family members and mitochondrial morphology is starting to be unraveled and similarly needs to be addressed using loss of function approaches. We expect years of exciting research to come.

Mitochondrial Ultrastructural Changes during Apoptosis

As occurs with Bax and Bak, also BH3 only proteins participate in cell death by modulation of mitochondrial fusion and fission. In addition, they display powerful activities on the mitochondrial inner shape. It has been shown that Bid is able to enlarge the narrow cristae junctions allowing the release of cytochrome *c* from the cristae to the IMS. This occurs via the disassembly of Opa1 complexes that normally maintain them tight. These complexes are formed by transmembrane and soluble isoforms of Opa1 and the precise mechanism by which Bid dismantles them is still a matter of discussion. First, it is still unclear which domains of Bid are required to induce it. In one report, the mobilization of cytochrome *c* release from the cristae seems not to depend on the BH3 domain of Bid and on Bax and Bak and it is inhibited by a GTPase functional form of Opa1.[124,129] However, another group reported that the Opa1-dependent cristae remodelling requires the presence, but not the activation, of Bax and Bak and it is blocked by mutations in the BH3 domain of Bid and Bim.[130] Further studies are necessary to verify if the Bid effect is due to a direct interaction with Opa1 complexes. An additional hypothesis is based on the finding that Bid is able to alter the membrane curvature in artificial lipid bilayers:[131] in mitochondria this effect on the membrane curvature could induce cristae remodelling without a direct interaction with Opa1. However, it would be difficult to reconcile a protein-independent effect with the ability of Opa1 to block the action of Bid on mitochondrial ultrastructure and release of cytochrome *c*. Clearly, our understanding of the process will benefit from the identification of the partners of Opa1 in healthy and apoptotic mitochondria.

Conclusion

Mitochondria are key players in cell death, whose involvement is modulated by the proteins of the BCL-2 family. Recently, we have learned that these proteins could display an intrinsic role in the regulation of mitochondrial shape, which is grossly altered early in the course of cell death. In the years to come we expect great advancements in the understanding of how mitochondrial shape changes regulate cell death and in our knowledge of the relative role of pro- and antiapoptotic members of the BCL-2 family in the regulation of mitochondrial morphology.

References

1. Vaux DL, Cory S, Adams JM. BCL-2 gene promotes haemopoietic cell survival and cooperates with c-myc to immortalize preB-cells. Nature 1988; 335:440-442.
2. Nunez G, Seto M, Seremetis S et al. Growth- and tumor-promoting effects of deregulated BCL2 in human B-lymphoblastoid cells. Proc Natl Acad Sci USA 1989; 86:4589-4593.
3. Hockenbery D, Nunez G, Milliman C et al. BCL-2 is an inner mitochondrial membrane protein that blocks programmed cell death. Nature 1990; 348:334-336.
4. McDonnell TJ, Nunez G, Platt FM et al. Deregulated BCL-2-immunoglobulin transgene expands a resting but responsive immunoglobulin M and D-expressing B-cell population. Mol Cell Biol 1990; 10:1901-1907.
5. Nunez G, London L, Hockenbery D et al. Deregulated BCL-2 gene expression selectively prolongs survival of growth factor-deprived hemopoietic cell lines. J Immunol 1990; 144:3602-3610.
6. Scaffidi C, Fulda S, Srinivasan A et al. Two CD95 (APO-1/Fas) signaling pathways. EMBO J 1998; 17:1675-1687.
7. Scaffidi C, Schmitz I, Zha J et al. Differential modulation of apoptosis sensitivity in CD95 type I and type II cells. J Biol Chem 1999; 274:22532-22538.
8. Medema JP, Scaffidi C, Kischkel FC et al. FLICE is activated by association with the CD95 death-inducing signaling complex (DISC). EMBO J 1997; 16:2794-2804.

9. Peter ME, Kischkel FC, Hellbardt S et al. CD95 (APO-1/Fas)-associating signalling proteins. Cell Death Differ 1996; 3:161-170.

10. Chinnaiyan AM, Tepper CG, Seldin MF et al. FADD/MORT1 is a common mediator of CD95 (Fas/APO-1) and tumor necrosis factor receptor-induced apoptosis. J Biol Chem 1996; 271:4961-4965.

11. Gross A, Yin XM, Wang K et al. Caspase cleaved BID targets mitochondria and is required for cytochrome c release, while BCL-XL prevents this release but not tumor necrosis factor-R1/Fas death. J Biol Chem 1999; 274:1156-1163.

12. Luo X, Budihardjo I, Zou H et al. Bid, a Bcl2 interacting protein, mediates cytochrome c release from mitochondria in response to activation of cell surface death receptors. Cell 1998; 94:481-490.

13. Yin XM, Wang K, Gross A et al. Bid-deficient mice are resistant to Fas-induced hepatocellular apoptosis. Nature 1999; 400:886-891.

14. Rich T, Allen RL, Wyllie AH. Defying death after DNA damage. Nature 2000; 407:777-783.

15. Blank M, Shiloh Y. Programs for cell death: apoptosis is only one way to go. Cell Cycle 2007; 6:686-695.

16. Roos WP, Kaina B. DNA damage-induced cell death by apoptosis. Trends Mol Med 2006; 12:440-450.

17. Gottlieb E, Armour SM, Thompson CB. Mitochondrial respiratory control is lost during growth factor deprivation. Proc Natl Acad Sci USA 2002; 99:12801-12806.

18. Saikumar P, Dong Z, Patel Y et al. Role of hypoxia-induced Bax translocation and cytochrome c release in reoxygenation injury. Oncogene 1998; 17:3401-3415.

19. Danial NN, Korsmeyer SJ. Cell death: critical control points. Cell 2004; 116:205-219.

20. Oltvai ZN, Milliman CL, Korsmeyer SJ. BCL-2 heterodimerizes in vivo with a conserved homolog, Bax, that accelerates programmed cell death. Cell 1993; 74:609-619.

21. Korsmeyer SJ, Shutter JR, Veis DJ et al. BCL-2/Bax: a rheostat that regulates an anti-oxidant pathway and cell death. Semin Cancer Biol 1993; 4:327-332.

22. Matsuyama S, Schendel SL, Xie Z et al. Cytoprotection by BCL-2 requires the pore-forming alpha5 and alpha6 helices. J Biol Chem 1998; 273:30995-31001.

23. Minn AJ, Velez P, Schendel SL et al. Bcl-x(L) forms an ion channel in synthetic lipid membranes. Nature 1997; 385:353-357.

24. Schendel SL, Xie Z, Montal MO et al. Channel formation by antiapoptotic protein BCL-2. Proc Natl Acad Sci USA 1997; 94:5113-5118.

25. Heimlich G, McKinnon AD, Bernardo K et al. Bax-induced cytochrome c release from mitochondria depends on alpha-helices-5 and -6. Biochem J 2004; 378:247-255.

26. Muchmore SW, Sattler M, Liang H et al. X-ray and NMR structure of human Bcl-xL, an inhibitor of programmed cell death. Nature 1996; 381:335-341.

27. Huang DC, Strasser A. BH3-Only proteins-essential initiators of apoptotic cell death. Cell 2000; 103:839-842.

28. Lutz RJ. Role of the BH3 (BCL-2 homology 3) domain in the regulation of apoptosis and BCL-2-related proteins. Biochem Soc Trans 2000; 28:51-56.

29. Nouraini S, Six E, Matsuyama S et al. The putative pore-forming domain of Bax regulates mitochondrial localization and interaction with Bcl-X(L). Mol Cell Biol 2000; 20:1604-1615.

30. Borner C. The BCL-2 protein family: sensors and checkpoints for life-or-death decisions. Mol Immunol 2003; 39:615-647.

31. Petros AM, Dinges J, Augeri DJ et al. Discovery of a potent inhibitor of the antiapoptotic protein Bcl-xL from NMR and parallel synthesis. J Med Chem 2006; 49:656-663.

32. Petros AM, Olejniczak ET, Fesik SW. Structural biology of the BCL-2 family of proteins. Biochim Biophys Acta 2004; 1644:83-94.

33. Petros AM, Medek A, Nettesheim DG et al. Solution structure of the antiapoptotic protein bcl-2. Proc Natl Acad Sci USA 2001; 98:3012-3017.

34. Petros AM, Nettesheim DG, Wang Y et al. Rationale for Bcl-xL/Bad peptide complex formation from structure, mutagenesis and biophysical studies. Protein Sci 2000; 9:2528-2534.

35. Gratiot-Deans J, Merino R, Nunez G et al. BCL-2 expression during T-cell development: early loss and late return occur at specific stages of commitment to differentiation and survival. Proc Natl Acad Sci USA 1994; 91:10685-10689.

36. Gratiot-Deans J, Ding L, Turka LA et al. BCL-2 proto-oncogene expression during human T-cell development. Evidence for biphasic regulation. J Immunol 1993; 151:83-91.

37. Chao DT, Korsmeyer SJ. BCL-XL-regulated apoptosis in T-cell development. Int Immunol 1997; 9:1375-1384.

38. Chao DT, Linette GP, Boise LH et al. BCL-XL and BCL-2 repress a common pathway of cell death. J Exp Med 1995; 182:821-828.

39. Moller C, Karlberg M, Abrink M et al. BCL-2 and BCL-XL are indispensable for the late phase of mast cell development from mouse embryonic stem cells. Exp Hematol 2007; 35:385-393.

40. Kamada S, Shimono A, Shinto Y et al. BCL-2 deficiency in mice leads to pleiotropic abnormalities: accelerated lymphoid cell death in thymus and spleen, polycystic kidney, hair hypopigmentation and distorted small intestine. Cancer Res 1995; 55:354-359.

41. Veis DJ, Sorenson CM, Shutter JR et al. BCL-2-deficient mice demonstrate fulminant lymphoid apoptosis, polycystic kidneys and hypopigmented hair. Cell 1993; 75:229-240.

42. Nakayama K, Nakayama K, Negishi I et al. Targeted disruption of BCL-2 alpha beta in mice: occurrence of gray hair, polycystic kidney disease and lymphocytopenia. Proc Natl Acad Sci USA 1994; 91:3700-3704.

43. Deaciuc I, D'Souza N, Nikolova-Karakashian M et al. The regulation of Fas (CD95/Apo-1)-mediated liver apoptosis in Kupffer cell-depleted mice. Hepatol Res 2002; 24:192.

44. Gibson L, Holmgreen SP, Huang DC et al. BCL-2, a novel member of the bcl-2 family, promotes cell survival. Oncogene 1996; 13:665-675.

45. Hamner S, Skoglosa Y, Lindholm D. Differential expression of bcl-w and bcl-x messenger RNA in the developing and adult rat nervous system. Neuroscience 1999; 91:673-684.

46. Dominov JA, Houlihan-Kawamoto CA, Swap CJ et al. Pro- and anti-apoptotic members of the BCL-2 family in skeletal muscle: a distinct role for BCL-2 in later stages of myogenesis. Dev Dyn 2001; 220:18-26.

47. O'Reilly LA, Print C, Hausmann G et al. Tissue expression and subcellular localization of the pro-survival molecule BCL-2. Cell Death Differ 2001; 8:486-494.

48. Yan W, Suominen J, Samson M et al. Involvement of BCL-2 family proteins in germ cell apoptosis during testicular development in the rat and pro-survival effect of stem cell factor on germ cells in vitro. Mol Cell Endocrinol 2000; 165:115-129.

49. Yan W, Samson M, Jegou B et al. BCL-2 forms complexes with Bax and Bak and elevated ratios of Bax/BCL-2 and Bak/BCL-2 correspond to spermatogonial and spermatocyte apoptosis in the testis. Mol Endocrinol 2000; 14:682-699.

50. Print CG, Loveland KL, Gibson L et al. Apoptosis regulator bcl-w is essential for spermatogenesis but appears otherwise redundant. Proc Natl Acad Sci USA 1998; 95:12424-12431.

51. Hershko A, Ciechanover A. The ubiquitin system. Annu Rev Biochem 1998; 67:425-479.

52. Cuconati A, Mukherjee C, Perez D et al. DNA damage response and MCL-1 destruction initiate apoptosis in adenovirus-infected cells. Genes Dev 2003; 17:2922-2932.

53. Opferman JT, Letai A, Beard C et al. Development and maintenance of B- and T-lymphocytes requires antiapoptotic MCL-1. Nature 2003; 426:671-676.

54. Chen L, Willis SN, Wei A et al. Differential targeting of prosurvival BCL-2 proteins by their BH3-only ligands allows complementary apoptotic function. Mol Cell 2005; 17:393-403.

55. Wang K, Gross A, Waksman G et al. Mutagenesis of the BH3 domain of BAX identifies residues critical for dimerization and killing. Mol Cell Biol 1998; 18:6083-6089.

56. Dzhagalov I, Dunkle A, He YW. The anti-apoptotic BCL-2 family member Mcl-1 promotes T-lymphocyte survival at multiple stages. J Immunol 2008; 181:521-528.

57. Rinkenberger JL, Horning S, Klocke B et al. Mcl-1 deficiency results in peri-implantation embryonic lethality. Genes Dev 2000; 14(1):23-714:23-27.

58. Opferman JT, Iwasaki H, Ong CC et al. Obligate role of anti-apoptotic MCL-1 in the survival of hematopoietic stem cells. Science 2005; 307:1101-1104.

59. Kaufmann SH, Karp JE, Svingen PA et al. Elevated expression of the apoptotic regulator Mcl-1 at the time of leukemic relapse. Blood 1998; 91:991-1000.

60. Zhou P, Levy NB, Xie H et al. MCL1 transgenic mice exhibit a high incidence of B-cell lymphoma manifested as a spectrum of histologic subtypes. Blood 2001; 97:3902-3909.

61. Ko JK, Lee MJ, Cho SH et al. Bfl-1S, a novel alternative splice variant of Bfl-1, localizes in the nucleus via its C-terminus and prevents cell death. Oncogene 2003; 22:2457-2465.

62. Somogyi RD, Wu Y, Orlofsky A et al. Transient expression of the BCL-2 family member, A1-a, results in nuclear localization and resistance to staurosporine-induced apoptosis. Cell Death Differ 2001; 8:785-793.

63. Grumont RJ, Rourke IJ, Gerondakis S. Rel-dependent induction of A1 transcription is required to protect B-cells from antigen receptor ligation-induced apoptosis. Genes Dev 1999; 13:400-411.

64. Kucharczak JF, Simmons MJ, Duckett CS et al. Constitutive proteasome-mediated turnover of Bfl-1/A1 and its processing in response to TNF receptor activation in FL5.12 pro-B-cells convert it into a prodeath factor. Cell Death Differ 2005; 12:1225-1239.

65. Zhang H, Cowan-Jacob SW, Simonen M et al. Structural basis of BFL-1 for its interaction with BAX and its anti-apoptotic action in mammalian and yeast cells. J Biol Chem 2000; 275:11092-11099.

66. Sedlak TW, Oltvai ZN, Yang E et al. Multiple BCL-2 family members demonstrate selective dimerizations with Bax. Proc Natl Acad Sci USA 1995; 92:7834-7838.
67. Hirotani M, Zhang Y, Fujita N et al. NH2-terminal BH4 domain of BCL-2 is functional for heterodimerization with Bax and inhibition of apoptosis. J Biol Chem 1999; 274:20415-20420.
68. Werner AB, de VE, Tait SW et al. BCL-2 family member Bfl-1/A1 sequesters truncated bid to inhibit is collaboration with pro-apoptotic Bak or Bax. J Biol Chem 2002; 277:22781-22788.
69. Suzuki M, Youle RJ, Tjandra N. Structure of Bax: coregulation of dimer formation and intracellular localization. Cell 2000; 103:645-654.
70. Nechushtan A, Smith CL, Hsu YT et al. Conformation of the Bax C-terminus regulates subcellular location and cell death. EMBO J 1999; 18:2330-2341.
71. Jurgensmeier JM, Xie Z, Deveraux Q et al. Bax directly induces release of cytochrome c from isolated mitochondria. Proc Natl Acad Sci USA 1998; 95:4997-5002.
72. Hsu YT, Wolter KG, Youle RJ. Cytosol-to-membrane redistribution of Bax and Bcl-X(L) during apoptosis. Proc Natl Acad Sci USA 1997; 94:3668-3672.
73. Annis MG, Soucie EL, Dlugosz PJ et al. Bax forms multispanning monomers that oligomerize to permeabilize membranes during apoptosis. EMBO J 2005; 24:2096-2103.
74. Antignani A, Youle RJ. How do Bax and Bak lead to permeabilization of the outer mitochondrial membrane? Curr Opin Cell Biol 2006; 18:685-689.
75. Chipuk JE, Green DR. How do BCL-2 proteins induce mitochondrial outer membrane permeabilization? Trends Cell Biol 2008; 18:157-164.
76. Schlesinger PH, Gross A, Yin XM et al. Comparison of the ion channel characteristics of proapoptotic BAX and antiapoptotic BCL-2. Proc Natl Acad Sci USA 1997; 94:11357-11362.
77. Antonsson B, Montessuit S, Lauper S et al. Bax oligomerization is required for channel-forming activity in liposomes and to trigger cytochrome c release from mitochondria. Biochem J 2000; 345(Pt 2):271-278.
78. Karbowski M, Lee YJ, Gaume B et al. Spatial and temporal association of Bax with mitochondrial fission sites, Drp1 and Mfn2 during apoptosis. J Cell Biol 2002; 159:931-938.
79. Han J, Flemington C, Houghton AB et al. Expression of bbc3, a pro-apoptotic BH3-only gene, is regulated by diverse cell death and survival signals. Proc Natl Acad Sci USA 2001; 98:11318-11323.
80. Wu X, Deng Y. Bax and BH3-domain-only proteins in p53-mediated apoptosis. Front Biosci 2002; 7:d151-d156.
81. Mandal M, Crusio KM, Meng F et al. Regulation of lymphocyte progenitor survival by the proapoptotic activities of Bim and Bid. Proc Natl Acad Sci USA 2008; 105:20840-20845.
82. Datta SR, Ranger AM, Lin MZ et al. Survival factor-mediated BAD phosphorylation raises the mitochondrial threshold for apoptosis. Dev Cell 2002; 3:631-643.
83. Datta SR, Katsov A, Hu L et al. 14-3-3 proteins and survival kinases cooperate to inactivate BAD by BH3 domain phosphorylation. Mol Cell 2000; 6:41-51.
84. Desagher S, Osen-Sand A, Nichols A et al. Bid-induced conformational change of Bax is responsible for mitochondrial cytochrome c release during apoptosis. J Cell Biol 1999; 144:891-901.
85. Li H, Zhu H, Xu CJ et al. Cleavage of BID by caspase 8 mediates the mitochondrial damage in the Fas pathway of apoptosis. Cell 1998; 94:491-501.
86. Letai A, Bassik MC, Walensky LD et al. Distinct BH3 domains either sensitize or activate mitochondrial apoptosis, serving as prototype cancer therapeutics. Cancer Cell 2002; 2:183-192.
87. Bereiter-Hahn J, Voth M. Dynamics of mitochondria in living cells: shape changes, dislocations, fusion and fission of mitochondria. Microsc Res Tech 1994; 27:198-219.
88. Frey TG, Mannella CA. The internal structure of mitochondria. Trends Biochem Sci 2000; 25:319-324.
89. Smirnova E, Griparic L, Shurland DL et al. Dynamin-related protein Drp1 is required for mitochondrial division in mammalian cells. Mol Biol Cell 2001; 12:2245-2256.
90. Pitts KR, Yoon Y, Krueger EW et al. The dynamin-like protein DLP1 is essential for normal distribution and morphology of the endoplasmic reticulum and mitochondria in mammalian cells. Mol Biol Cell 1999; 10:4403-4417.
91. James DI, Parone PA, Mattenberger Y et al. hFis1, a novel component of the mammalian mitochondrial fission machinery. J Biol Chem 2003; 278:36373-36379.
92. Cereghetti GM, Stangherlin A, Martins de BO et al. Dephosphorylation by calcineurin regulates translocation of Drp1 to mitochondria. Proc Natl Acad Sci USA 2008; 105:15803-15808.
93. Yoon Y, Krueger EW, Oswald BJ et al. The mitochondrial protein hFis1 regulates mitochondrial fission in mammalian cells through an interaction with the dynamin-like protein DLP1. Mol Cell Biol 2003; 23:5409-5420.
94. Santel A, Fuller MT. Control of mitochondrial morphology by a human mitofusin. J Cell Sci 2001; 114:867-874.

95. Legros F, Lombes A, Frachon P et al. Mitochondrial fusion in human cells is efficient, requires the inner membrane potential and is mediated by mitofusins. Mol Biol Cell 2002; 13:4343-4354.

96. Chen H, McCaffery JM, Chan DC. Mitochondrial Fusion Protects against Neurodegeneration in the Cerebellum. Cell 2007; 130:548-562.

97. Chen H, Detmer SA, Ewald AJ et al. Mitofusins Mfn1 and Mfn2 coordinately regulate mitochondrial fusion and are essential for embryonic development. J Cell Biol 2003; 160:189-200.

98. Akepati VR, Muller EC, Otto A et al. Characterization of OPA1 isoforms isolated from mouse tissues. J Neurochem 2008; 106:372-383.

99. Cipolat S, de Brito OM, Dal Zilio B et al. OPA1 requires mitofusin 1 to promote mitochondrial fusion. Proc Natl Acad Sci USA 2004; 101:15927-15932.

100. Cribbs JT, Strack S. Reversible phosphorylation of Drp1 by cyclic AMP-dependent protein kinase and calcineurin regulates mitochondrial fission and cell death. EMBO Rep 2007; 8:939-944.

101. Chang CR, Blackstone C. Cyclic AMP-dependent protein kinase phosphorylation of Drp1 regulates its GTPase activity and mitochondrial morphology. J Biol Chem 2007; 282:21583-21587.

102. Taguchi N, Ishihara N, Jofuku A et al. Mitotic phosphorylation of dynamin-related GTPase Drp1 participates in mitochondrial fission. J Biol Chem 2007; 282:11521-11529.

103. Harder Z, Zunino R, McBride H. Sumo1 conjugates mitochondrial substrates and participates in mitochondrial fission. Curr Biol 2004; 14:340-345.

104. Wasiak S, Zunino R, McBride HM. Bax/Bak promote sumoylation of DRP1 and its stable association with mitochondria during apoptotic cell death. J Cell Biol 2007; 177:439-450.

105. Anesti V, Scorrano L. The relationship between mitochondrial shape and function and the cytoskeleton. Biochim Biophys Acta 2006; 1757:692-699.

106. Martinou I, Desagher S, Eskes R et al. The release of cytochrome c from mitochondria during apoptosis of NGF-deprived sympathetic neurons is a reversible event. J Cell Biol 1999; 144:883-889.

107. Frank S, Gaume B, Bergmann-Leitner ES et al. The role of dynamin-related protein 1, a mediator of mitochondrial fission, in apoptosis. Dev Cell 2001; 1:515-525.

108. Karbowski M, Arnoult D, Chen H et al. Quantitation of mitochondrial dynamics by photolabeling of individual organelles shows that mitochondrial fusion is blocked during the Bax activation phase of apoptosis. J Cell Biol 2004; 164:493-499.

109. Jahani-Asl A, Cheung EC, Neuspiel M et al. Mitofusin 2 protects cerebellar granule neurons against injury induced cell death. J Biol Chem 2007; 282:23788-23798.

110. Barsoum MJ, Yuan H, Gerencser AA et al. Nitric oxide-induced mitochondrial fission is regulated by dynamin-related GTPases in neurons. EMBO J 2006; 25:3900-3911.

111. Leinninger GM, Backus C, Sastry AM et al. Mitochondria in DRG neurons undergo hyperglycemic mediated injury through Bim, Bax and the fission protein Drp1. Neurobiol Dis 2006; 23:11-22.

112. Yuan H, Gerencser AA, Liot G et al. Mitochondrial fission is an upstream and required event for bax foci formation in response to nitric oxide in cortical neurons. Cell Death Differ 2007; 14:462-471.

113. Dagda RK, Merrill RA, Cribbs JT et al. The spinocerebellar ataxia 12 gene product and protein phosphatase 2A regulatory subunit Bβ2 antagonizes neuronal survival by promoting mitochondrial fission. J Biol Chem 2008; 283:36241-36248.

114. Parra V, Eisner V, Chiong M et al. Changes in mitochondrial dynamics during ceramide-induced cardiomyocyte early apoptosis. Cardiovasc Res 2008; 77:387-397.

115. Yu T, Sheu SS, Robotham JL et al. Mitochondrial fission mediates high glucose-induced cell death through elevated production of reactive oxygen species. Cardiovasc Res 2008; 79:341-351.

116. Yu T, Robotham JL, Yoon Y. Increased production of reactive oxygen species in hyperglycemic conditions requires dynamic change of mitochondrial morphology. Proc Natl Acad Sci USA 2006; 103:2653-2658.

117. Parone PA, James DI, Da CS et al. Inhibiting the mitochondrial fission machinery does not prevent Bax/Bak-dependent apoptosis. Mol Cell Biol 2006; 26:7397-7408.

118. Estaquier J, Arnoult D. Inhibiting Drp1-mediated mitochondrial fission selectively prevents the release of cytochrome c during apoptosis. Cell Death Differ 2007; 14:1086-1094.

119. Cassidy-Stone A, Chipuk JE, Ingerman E et al. Chemical inhibition of the mitochondrial division dynamin reveals its role in Bax/Bak-dependent mitochondrial outer membrane permeabilization. Dev Cell 2008; 14:193-204.

120. Jagasia R, Grote P, Westermann B et al. DRP-1-mediated mitochondrial fragmentation during EGL-1-induced cell death in C. elegans. Nature 2005; 433:754-760.

121. Goyal G, Fell B, Sarin A et al. Role of mitochondrial remodeling in programmed cell death in Drosophila melanogaster. Dev Cell 2007; 12:807-816.

122. Sheridan C, Delivani P, Cullen SP et al. Bax- or Bak-induced mitochondrial fission can be uncoupled from cytochrome C release. Mol Cell 2008; 31:570-585.

123. Sugioka R, Shimizu S, Tsujimoto Y. Fzo1, a protein involved in mitochondrial fusion, inhibits apoptosis. J Biol Chem 2004; 279:52726-52734.
124. Frezza C, Cipolat S, Martins dB et al. OPA1 controls apoptotic cristae remodeling independently from mitochondrial fusion. Cell 2006; 126:177-189.
125. Men X, Wang H, Li M et al. Dynamin-related protein 1 mediates high glucose induced pancreatic beta cell apoptosis. Int J Biochem Cell Biol 2009; 41:879-890.
126. Fannjiang Y, Cheng WC, Lee SJ et al. Mitochondrial fission proteins regulate programmed cell death in yeast. Genes Dev 2004; 18:2785-2797.
127. Delivani P, Adrain C, Taylor RC et al. Role for CED-9 and Egl-1 as regulators of mitochondrial fission and fusion dynamics. Mol Cell 2006; 21:761-773.
128. Karbowski M, Norris KL, Cleland MM et al. Role of Bax and Bak in mitochondrial morphogenesis. Nature 2006; 443:658-662.
129. Scorrano L, Ashiya M, Buttle K et al. A distinct pathway remodels mitochondrial cristae and mobilizes cytochrome c during apoptosis. Dev Cell 2002; 2:55-67.
130. Yamaguchi R, Lartigue L, Perkins G et al. Opa1-mediated cristae opening is Bax/Bak and BH3 dependent, required for apoptosis and independent of Bak oligomerization. Mol Cell 2008; 31:557-569.
131. Epand RF, Martinou JC, Fornallaz-Mulhauser M et al. The apoptotic protein tBid promotes leakage by altering membrane curvature. J Biol Chem 2002; 277:32632-32639.

CHAPTER 7

Noncanonical Functions of BCL-2 Proteins in the Nervous System

Heather M. Lamb and J. Marie Hardwick*

Abstract

BCL-2 family proteins form heterodimers or homo-oligomers to inhibit or induce apoptotic cell death, respectively. They often relocalize from the cytoplasm to mitochondria to carry out these functions. The traditional model is that in healthy cells, anti-death family members hold pro-death BCL-2 family members in check. Upon receiving a death stimulus, another set of proteins (BH3-only proteins) inactivate the protective BCL-2 proteins, forcing them to release their pro-death partners that are subsequently triggered to oligomerize and porate the mitochondrial outer membrane leading to cell death. In support of this traditional view, there is a preponderance of supporting evidence derived from the study of events that occur following treatment of cells with a death stimulus. Knockout and mutant mice also exhibit many developmental and treatment-induced phenotypes consistent with this model of antagonism between BCL-2 family proteins. Emphasis is logically placed on those phenotypes that support the model. However, this working model of BCL-2 family interactions has become so engrained that alternative, potentially valid interpretations are sometimes dismissed. Therefore, it is useful to consider the evidence that seems contrary to accepted models. In particular, the analysis of BCL-2 family functions in the nervous system has revealed unexpected outcomes that can serve to further stimulate critical probing of the yet unknown biochemical functions of BCL-2 proteins.

Introduction

BCL-2 family proteins are best known for their roles in regulating apoptotic cell death during the final 20 minutes of the life of a cell. These final minutes are the focus of intense effort to understand the underlying molecular details. However, the functions of BCL-2 proteins in healthy cells are seldom investigated. We assume that their functions in healthy cells are to prepare for their functions after a death stimulus, but this has not been unequivocally confirmed. A long-held assumption is that pro-death BCL-2 proteins exist in an inactive state or are locked into inactive complexes of paired anti- and pro-death heterodimers prior to a death stimulus. However, there is also evidence that BCL-2 family proteins are actively engaged in promoting functions important to healthy cells. For example, BCL-2 proteins appear to be important for the synaptic activity of healthy adult neurons, long before the activation of a death pathway. If most BCL-2 family proteins are actively engaged in promoting normal physiological functions in healthy cells, these mechanisms could be distinct from those following a death stimulus (Fig. 1). Distinguishing healthy-cell functions from dying-cell functions of BCL-2 proteins could be tricky, analogous to the complicated task of distinguishing between the functions of cytochrome c. Deletion of cytochrome c is lethal,

*Corresponding Author: J Marie Hardwick—Johns Hopkins (JHSPH), Department of Molec Microb and Immunol, 615 N Wolfe St., Baltimore, Maryland 21205, USA.
Email: hardwick@jhu.edu

BCL-2 Protein Family: Essential Regulators of Cell Death, edited by Claudio Hetz.
©2010 Landes Bioscience and Springer Science+Business Media.

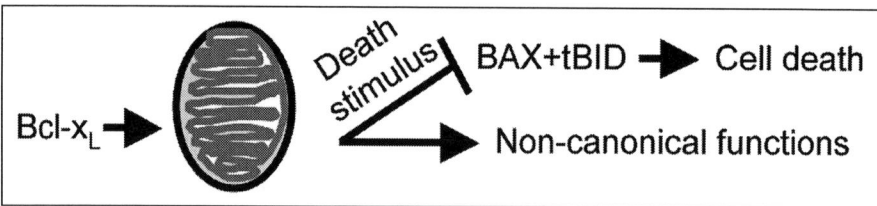

Figure 1. Bcl-x$_L$, Bak and other BCL-2 family proteins may have normal physiological (noncanonical) functions in healthy cells that are distinct from their roles following a death stimulus. Bcl-x$_L$ and Bak can regulate synaptic activity in healthy neurons.

but we can avoid the mistaken conclusion that cytochrome *c* is anti-apoptotic because we know it has an established pro-survival function in mitochondrial respiration. In fact, cytochrome *c* is pro-apoptotic following a death stimulus when it catalyzes caspase activation. Could we be making analogous errors in the interpretation of BCL-2 family functions? Especially when studying the nervous system, some of the evidence regarding BCL-2 family proteins does not seem to fit or even challenges the currently accepted models of their mechanisms of action. However, results that seemingly conflict with the model are often not reported or further pursued. It is important to keep in mind that a detailed biochemical mechanism that is consistent with the available 3-dimensional structures of BCL-2 proteins and that explains their cell survival function is currently lacking.

Extending the Traditional View of BCL-2 Family Proteins

Before analyzing the unexpected findings for BCL-2 family proteins, it is useful to briefly consider the traditional view of BCL-2 proteins. The gene for the prototype BCL-2 protein was first identified as a probable oncogene at the chromosome translocation breakpoints that are characteristic of follicular lymphomas.[1-4] Coinciding with the identification of human BCL-2, the *C. elegans* homologue CED-9 was shown to be required for cell survival during development and to inhibit caspase-dependent cell death.[5,6] Significant amino acid sequence similarity was noted between BCL-2 and the early viral gene product of unknown function, BHRF1, encoded by Epstein-Barr virus, a herpesvirus and a causal agent of B-cell lymphomas and other human cancers.[4] But rather than increasing cell growth rates like the oncogenes identified previously, BCL-2 and BHRF1 were shown to keep cells alive under conditions where they would otherwise die.[7,8] Subsequently, ~10 additional human BCL-2-related proteins were confirmed or predicted to adopt the characteristic 3-dimensional protein structure of the BCL-2 family.[9,10] Despite their shared structures, at least four of these family members kill cells rather than inhibit cell death (Bax, Bak, Bok, Bid).

It is widely thought that the anti-death family members (BCL-2, Bcl-x$_L$, Mcl-1, BCL-2, A1/Bfl1) protect cells from death by directly binding and inhibiting their pro-death family members.[11] The pro-death function of Bax and Bak is derived from their ability to form homo-oligomers in the mitochondrial outer membrane, a process triggered by cleaved/truncated Bid (tBid).[12,13] It is widely assumed that the membrane-associated Bax/Bak homo-oligomer forms the conduit, or otherwise directly mediates the passage of cytochrome *c* across the outer membrane into the cytoplasm, where cytochrome *c* catalyzes the activation of apoptotic caspases leading to death.[13-15] Based on recently solved 3-dimensional structures of several viral proteins that lack obvious amino acid sequence similarity to BCL-2, including 9 ORFs encoded by a single virus genome,[16,17] there are potentially many other yet unidentified BCL-2-like proteins. Interestingly, the BCL-2 sequence homologues encoded by *Drosophila* (Debcl and Buffy) and the BALF1 protein of Epstein-Barr virus, as well as several vaccinia virus structural mimics of BCL-2, have limited or no known roles in regulating apoptosis.[17-19]

Bid has traditionally been classified with a second pro-apoptotic subgroup of the BCL-2 family, a set of ~8 human proteins referred to as BH3-only proteins (tBid, Bad, Bik, Bim, Bmf, Hrk,

Noxa, Puma). BH3-only proteins have only one of the four BCL-2 homology (BH) domains/ motifs and lack overall amino acid sequence similarity to each other or to (multi-domain) BCL-2 proteins. The BH3 domain of BH3-only proteins can insert into a binding groove on specific anti-death BCL-2 family members, inactivating their anti-death function. Although many questions and controversies remain, BH3-only proteins are suggested to be sensors of cellular damage/stress, to which they respond by binding and inactivating anti-apoptotic BCL-2 family members.[20] Only a subset of these BH3-only proteins (e.g., tBid) has the capacity to directly stimulate oligomerization of Bax and Bak apparently in a hit-and-run mechanism.[21-23] With the exception of Bid, which possesses a bona fide BCL-2-like 3-dimensional structure, attempts have failed thus far to determine the 3-dimensional fold for BH3-only proteins. Except for their BH3 domains, BH3-only proteins appear to be unstructured when bound to their BCL-2 protein partners.[24,25] Thus, the relatedness of these seven BH3-only proteins to the BCL-2 family remains in question. Nevertheless, peptides and small molecules that mimic BH3 domains appear likely to be useful therapeutics for the treatment of human cancers.[26-28] These molecules insert into the binding groove of anti-death BCL-2 family proteins and kill cells in a Bax/Bak-dependent manner, providing compelling evidence that endogenous BH3-only proteins may act in a similar manner. However, off-target effects have also been observed, as a stapled peptide mimetic of the Bad BH3 domain appears to target glucokinase rather than Bcl-x_L.[29] The NMR structure of a Bim version of the BH3 stapled peptide was recently solved, revealing a new binding site on the opposite side of Bax from the expected binding groove typical of anti-death BCL-2 proteins.[30] Extrapolating from in vitro BH3 binding studies to all of the relevant intracellular targets is clearly a challenge for the future.

In addition to the traditional group of BH3-only proteins, several clearly unrelated proteins (e.g., Beclin and Mule) have been shown to contain a BH3-like motif that also binds the groove on BCL-2 proteins to inactivate their functions.[31,32] In addition, BCL-2 family proteins have also been reported to directly interact with a wide variety of other cellular proteins, including transcription factors, channel proteins, cytoskeletal components, autophagy regulators, GTPases, kinases and numerous others.[33-38] However, a central theme that links these interactions to a biochemical function of BCL-2 proteins has not been delineated.

Challenging the Traditional BCL-2 Family Model

There are a number of intriguing findings that challenge the general assumptions about the BCL-2 family summarized above with regard to the interaction of anti- and pro-death BCL-2 family proteins. The first wedge was driven into the Bax-BCL-2 antagonism/rheostat model with the generation of point mutants of BCL-2 and Bcl-x_L that are unable to bind Bax and Bak but retain their anti-apoptotic function.[39] (unpublished data). This raised the possibility that BCL-2/ Bcl-x_L might have Bax/Bak-independent activities.

Point mutations that interfere with the ability of a BCL-2 family member to bind one but not the other of its partners has lead to the conclusion that Bak is not normally sequestered by anti-death BCL-2 proteins, but by VDAC2 (voltage-dependent anion channel), the abundant metabolite transporter on the outer mitochondrial membrane.[40,41] The evidence suggests that specific BH3-only proteins can displace this interaction between VDAC2 and Bak in healthy cells, leading to cell death.

Detergent-induced heterodimer formation between BCL-2/Bcl-x_L proteins and their pro-death Bax/Bak homologues during extract preparation was shown to explain why these interactions are so easily detected in co-immunoprecipitation experiments, further challenging the model.[42] Though BCL-2-Bax interactions could be avoided by using CHAPS instead of NP-40, it raised awareness regarding potential artifacts occurring in cell extracts that are not reflective of endogenous states. However, it could be argued that detergents partially mimic the effects of endogenous lipid membranes on BCL-2 protein conformations. Nevertheless, the fact that different detergents reveal or conceal different BCL-2 family heterodimers and the observation that lipid membranes are required to observe interactions between tBid and Bax (an interaction disrupted by CHAPS) further complicates the interpretation of such experiments.[23]

An alternative view is that BCL-2 family proteins act by remodeling membranes, which in turn alters the functions of other membrane-associated factors.[43,44] If indeed BCL-2 inhibits cell death by altering membrane structures independently of any direct interactions with other proteins, this could potentially explain the striking finding that BCL-2 can inhibit cell death in yeast and plants despite the lack of obvious BCL-2-related sequences in these species.[45,46]

BCL-2 Family Proteins in the Nervous System

The expression patterns of BCL-2 proteins and analyses of their corresponding knockout mice have revealed important roles for BCL-2 family proteins and BH3-only proteins in the nervous system, both during development and in several disease models. However, some of these studies challenge the current model for BCL-2 family mechanisms of action described above. Expression of BCL-2 family proteins is developmentally regulated in the nervous system, with BCL-2, Bcl-x$_L$, Mcl-1, Bax, Bak and Bid expressed during embryogenesis.[47-50] In adult mouse brain, Bcl-x$_L$ and BCL-2 appear to be the primary BCL-2 family molecules expressed in neurons (Krajewska et al., Neoplasia 4:2 pp129-140 (2002)). Bid expression in adult neurons has been described, but has yet to be confirmed with appropriate controls. Bax and Bak expression also occurs in nonneuronal cell types of the brain and their expression declines significantly through adulthood.[47] Although Bak is suggested to be expressed only in nonneuronal cells of the nervous system, an alternatively spliced BH3-only isoform of Bak (N-Bak) was reported in postnatal cortical and cerebellar granule cells and promotes cell death by directly binding Bcl-x$_L$.[52] A constitutively active splice variant of Bax, Bax beta, which has a BH3 domain and distinct C-terminus, was recently identified in a variety of cell lines.[53] These and other reports of alternatively spliced BCL-2 family and BH3-only proteins raise awareness of potentially overlooked variants.

Bax/Bak double knockout mice have a striking excess of neuroprogenitor cells and this phenotype is mild or undetectable in the single knockouts.[54] Bax-deficiency protects specific neuronal subtypes from various insults, but is without effect in other scenarios.[54-56] Deletion of Bax or Bak can have opposite to expected effects as discussed further below. BCL-2 deficiency appears not to affect neuronal numbers in the cerebellum,[57] but has been implicated in hippocampal neuron protection in a seizure model.[58] In contrast, conditional deletion of Mcl-1 revealed a role for Mcl-1 in both neuroprogenitor cell survival and in DNA damage-induced neuronal death.[50] Although BCL-2-deficient mice appear to have relatively normal development of the nervous system, BCL-2 is important for the postnatal maintenance of specific neuronal subsets, including facial motor neurons and dorsal root ganglia.[59-61]

Bcl-x$_L$ is essential for embryonic development, as gene deletion causes lethality at embryonic day E12.5-13.5 with death of immature hematopoetic cells and neurons.[62] However, tissue-specific deletion of Bcl-x$_L$ in cortical, hippocampal or catecholaminergic neurons produces viable animals with reductions in the corresponding neuron subtypes.[63,64] Healthy neurons prepared from *bcl-x* conditional knockout mice can be maintained in culture, though they are more susceptible to the stress of explantation and in vitro culture conditions.[65] Some developmental and stimulus-induced neuronal death that occurs in *bcl-x*-deficient embryos and derived cells can be delayed in *bcl-x/bax* double knockouts, consistent with functional interactions between Bax and Bcl-x$_L$ in the brain.[66,67] However, the striking finding from these and other studies is the general lack of a clear genetic interaction between anti- and pro-death BCL-2 family members in the nervous system of mice with double deletions.[66,68] For example, attempts to rescue long-term neuronal survival (and embryonic lethality) of *bcl-x*-deficient mice by deletion of pro-apoptotic BCL-2 family proteins, BH3-only proteins or unrelated death factors have not been successful thus far.[69-71] Similar results were observed in erythropoiesis where Bax-deficiency failed to inhibit the late erythroid cell death that occurs in *bcl-x* conditional knockouts.[72] However, the loss of platelets due to Bcl-x$_L$ genome mutations appears to be mediated in part by Bak, but not Bax.[73,74] Despite evidence that Bak promotes death of neuroprogenitors, full-length Bak protein is not expressed at detectable levels in neurons. Thus, it remains possible that very low expression in neurons or the effects of Bak in nonneuronal cells in the brain could potentiate Bcl-x_L-deficiency.

Results with double knockout mice have not strongly supported a direct interaction of anti- and pro-death (multi-domain) BCL-2 family proteins in the nervous system. However, the presence of multiple BCL-2 family members, the variable timing of their expression patterns and an incomplete exploration of all combinations of BCL-2 proteins of genetic studies (e.g., Bok) further complicate the interpretation of these animal experiments. Still, the current model of BCL-2 family protein interactions may not be sufficient to explain the functions of the (multi-domain) BCL-2 family proteins. Expression patterns of anti- and pro-death BCL-2 family proteins in the adult nervous system are asynchronous. This imbalance could reflect purposeful adjustments to favor the function of one over the other, but could also suggest independent functions. The anti-apoptotic protein Bcl-x_L is abundantly expressed in neurons of the adult brain when the levels of pro-apoptotic Bax and Bak are down regulated or possibly absent in neurons. Definitive proof of direct or functional interactions between BCL-2/Bcl-x_L and Bax/Bak type proteins in the nervous system is lacking thus far. This is in striking contrast to overexpression of BCL-2/Bcl-x_L and Bax/Bak type proteins in cultured cells, where they efficiently counteract each other's effects on cell survival.

Reports are mixed on the role of BH3-only protein Bim in neuron development, likely due to differences between neuronal subtypes. Although deletion of Bim in Bcl-x_L[+/-] adult mice partially rescues death of some cell types, no rescue of neuronal death was detected.[71] Several different BH3-only proteins can be induced upon neuronal injury and may play important roles, but knockout mice have confirmed functional relevance for very few of these. Bim-deficient mice were reported to be significantly protected in a transgenic SOD1[G93A] mouse model of ALS.[75,76] Bim-deficiency also protects against optic nerve axotomy-induced cell death in a retinal degeneration model, but the delay in cell death without Bim did not ultimately change neuron survival rates.[75,77] This is in contrast to the striking rescue of kidney development in BCL-2-deficient mice by simultaneous deletion of only one copy of the BH3-only protein Bim.[78]

When Anti-Death BCL-2 Proteins Become Pro-Death

As summarized above, the anti-death functions of endogenous BCL-2, Bcl-x_L, BCL-2 and Mcl-1 in neurons have been confirmed in multiple studies of knockout and conditional knockout mice. However, there are a few examples where anti-apoptotic BCL-2 family proteins can exhibit profound pro-death activities. Using Sindbis virus as a vector to overexpress BCL-2 family proteins predominantly in neurons of the brain and spinal cord of infected mice,[79,80] overexpression of Bcl-x_L dramatically enhances neuronal death and mouse mortality compared to several control viruses (unpublished). This is in sharp contrast to the same experiment performed with BCL-2, where infection of mice with Sindbis virus encoding BCL-2 results in significant protection of neurons and increased mouse survival significantly compared to control viruses.[81,82] Consistent with this result, BCL-2 knockout mice are more susceptible to a fatal Sindbis virus infection.[83] However, we have generated several N-terminal mutants of BCL-2 that can turn deadly in the same assay system (unpublished).

The mechanisms that mediate this dramatic conversion of anti-apoptotic BCL-2 proteins to pro-death function in neurons of the brain are unknown, but several hypotheses are consistent with these findings. One of these involves proteolytic processing of BCL-2 and Bcl-x_L. Analogous to Bid, which needs to be cleaved to efficiently activate Bax/Bak to kill, anti-apoptotic proteins BCL-2 and Bcl-x_L can be cleaved by caspases and other proteases to release C-terminal fragments capable of triggering cytochrome *c* release and cell death.[84-88] Alternatively, the ability of Bcl-x_L to stimulate synaptic activity, could explain its toxic effect, for example if Bcl-x_L triggers excitotoxicity (see below). A pro-death function of anti-apoptotic BCL-2 proteins could potentially explain why they contain a BH3 motif capable of triggering cell death in some cases, but not others.[89]

The BH3-only protein Bad is considered to be a pro-death protein, but often fails to kill cells when overexpressed. When Bad does kill cells, it appears to be dependent on proteolytic cleavage at one of three caspase cleavage sites encoded within a single exon.[90,91] Interestingly, the specific death stimulus applied determines which one of the caspase cleavage sites is important for cell death. Furthermore, overexpression of Bad in mouse brains and cultured neurons can potently protect from cell death and Sindbis virus-induced mouse mortality.[90] Other studies of endogenous and overexpressed Bad are consistent with a normal cellular pro-survival function,[91,92] though some of the evidence has to be inferred from the data presented. A role for Bad in glucose metabolism further testifies to the importance of Bad, in this case by preventing diabetes.[29] The death function of Bax may also be enhanced by proteolysis[85,93] and several other BCL-2-related proteins and BH3-only proteins have been reported to be cleaved by caspases and other proteases in a topologically similar manner.[94-99]

Alternatively, anti-death proteins may adopt pro-death activities by altering the conformation of the N-terminus of anti-death BCL-2 proteins, potentially mimicking proteolytic cleavage. Thus, nonproteolytic events may be sufficient to convert anti-death proteins into pro-death factors.[100] In the case of Bad, proteolytic clipping is not always sufficient to convert Bad into a pro-death factor in immature neurons, whereas phosphorylation events are apparently also required.[90] Although outside the nervous system, it is interesting that the nuclear orphan receptor Nur77, a Nur77-derived peptide and small molecule mimics of Nur77 were reported to bind the N-terminal region of BCL-2 and Bcl-B and convert them into pro-death molecules.[101-103] The mechanism appears to be a major conformational change that involves both the N- and C-termini of BCL-2/Bcl-B, serving as a type of molecular switch.

Does the conversion of anti-apoptotic BCL-2 proteins into pro-death factors occur during physiological cell death? The difficulty of distinguishing between the loss of anti-death activity and the conversion to pro-death activity makes this task difficult. However, Nur77 was recently reported to translocate to mitochondria, where it was suggested to convert BCL-2 to a BH3-exposed killer protein during negative selection of autoreactive T-cells in a mouse model.[104] These observations suggest new strategies for treating a variety of cancers where tumor cell survival is mediated by anti-apoptotic BCL-2 family proteins.

In addition to the obvious conversion of Bcl-x$_L$ into a killer in Sindbis virus infections of neurons in the brain, this conversion function may also operate in neurons during hypoxia-induced changes in neuronal excitability. There is considerable evidence that the small molecule inhibitor, ABT-737, designed to occupy the BH3-binding groove of Bcl-x$_L$, kills tumor cells by binding and inactivating anti-death BCL-2 family proteins, except for Mcl-1.[26] Quite unexpectedly, ABT-737 protects neurons from hypoxia-induced synaptic decline and from increased mitochondrial permeability in a squid model. This finding implies that ABT-737 can also inactivate the pro-apoptotic form of BCL-2 and/or Bcl-x$_L$ and perhaps also BCL-2.[105] This conclusion is supported by the observation that ABT-737 can significantly inhibit synaptic rundown induced by injection of an engineered caspase cleavage fragment of Bcl-x$_L$ into the presynaptic terminus.[105]

When Pro-Death BCL-2 Proteins Protect Neurons

Infection of newborn mice with Sindbis virus, which predominantly infects neurons in the brain and spinal cord, induces neuronal death and mortality with encephalitis.[79,106-109] Sindbis virus has been used for the past 50 years to model closely related human encephalitic viruses that are mosquito-borne pathogens and pose a significant health risk. The Sindbis virus RNA genome was engineered to carry a copy of BCL-2.[110] As briefly mentioned above, infection of mice with this BCL-2-expressing virus results in significantly less neuronal death and reduced mortality compared to control Sindbis viruses.[81,82] We also investigated the possibility that Bax and Bak could be important virulence determinants. To our initial surprise, the opposite was true. That is, Sindbis viruses encoding either Bax or Bak dramatically protected mice from neuronal death and mortality and more potently than BCL-2.[80,111] Consistent with these unexpected findings using overexpressed proteins, juvenile Bax or Bak knockout mice and their derived tissues, are

Anti, pro-death or neutral functions of endogenous BAK and BAX									
Mouse model	Encephalomyelitis					Seizure		Parkinson disease	Stroke
Experimental method	Sindbis virus					Kainate/ glutamate		MPTP Toxicity	MCAO
Target neurons	Hippo-campal		Spinal cord		Mortality	Hippo-campal		Dopami-nergic	Cortical/ striatal
Age (weeks)	0.5	4	0.5	4	0.5	4	1	4	8-11	9-12
BAK	Anti	Anti	Anti	Pro	Anti	Pro	Anti	Anti	Anti	Pro
BAX	Anti	Anti	Anti	Neu	Anti	Anti	?	?	Pro	Pro

Figure 2. Bak and Bax knockout mice analyzed in several mouse models of human neurological disorders. Endogenous Bak and Bax can function as pro- or anti-death molecules depending on the age, target tissue and death stimulus as previously reported.[111] Reprinted from Fannjiang Yet al. Dev Cell 2003; 4:575-585, with permission from Elsevier.

significantly more susceptible to cell death and virus-induced mortality.[80,111] Even more remarkable, reconstitution of these knockout mice by infecting with Sindbis viruses expressing Bax or Bak rescues mice from a lethal infection.

Although contrary to current dogma, *bak⁻/⁻* mice were significantly more susceptible to hippocampal neuron death when injected with kainate to induce acute seizure activity, and Bak protects against subsequent neuronal death following a seizure.[111] Bak also plays a role in preventing excitotoxic injury, as slice cultures from Bak-deficient mice were more susceptible to neuronal death associated with kainate, NMDA, or glutamate treatment.[111] An extensive analysis of Bak knockout mice using a variety of assays related to brain disorders (stroke, Parkinson's disease and epilepsy) revealed that Bak can have opposite functions depending on the affected neuronal subtype, developmental stage and the specific death stimulus[111] (Fig. 2). While Bak and Bax protect against cell death in acute and cultured brain slices, the same neurons are highly susceptible to Bax- and Bak-induced death when these neurons are dissociated and plated. These findings suggest that the connectivity of neurons in the brain strongly influences the function of Bax/Bak following an insult.

Bax was also reported to mediate the survival of various sensory and parasympathetic neuronal cultures deprived of neurotrophic factors.[112,113] However, this pro-survival function of Bax was not observed in the superior cervical ganglion,[112] consistent with cell-type specific functions in neurons. The protective effects of endogenous and overexpressed Bak in neurons seem to be in conflict with the frequent failure to detect expression of Bak in neurons. It is important to keep in mind that in situ techniques are relatively insensitive. Furthermore, defects in astrocytes and glial cells are often first recognized by the loss of neurons, as multiple cell types in the brain support neurons. Whether the pro-death function of Bak that contributes to the significant increase in neuroprogenitor cell survival observed in the Bax/Bak double knockout is due to direct effects of Bak in neurons, glia or both is not clear.[54] However, we assume that the protective effects of Bak (as well as BCL-2, Bax and Bad) overexpressed by the Sindbis virus vector in mouse brains is due to increased neuronal expression due to the known tropism of this virus.[109,111]

These reverse functions may be evolutionarily conserved. Knockdown of *Drosophila* Debcl/Bok/Drob-1, an orthologue of mammalian Bok/Bax/Bak, in neurons was found to protect against neurodegeneration induced by polyglutamine[114] and to alter intracellular ATP levels and mitochondrial membrane potential in response to complex I and complex II inhibitors, suggesting a role for Debcl/Drob-1 in maintaining mitochondrial homeostasis. However, evidence for a general role of Debcl and the anti-death BCL-2-related factor Buffy in regulating cell death in *Drosophila* analogous to that in worms is strikingly lacking.[18,115,116]

Protection Predicts New Functions with a Seizure

The pro-death function of BCL-2 could be related to the function of Bax and vise versa, but it is also possible that both anti- and pro-death family members have a yet undiscovered, shared biochemical function. The hypothesis that BCL-2 family members possess alternative noncanonical functions in neurons stemmed from observations of Bak knockout mice. Mice lacking Bak appear to be developmentally normal by gross and histological analyses.[54] However, *bak*$^{-/-}$ mice had more severe seizures following treatment with kainate[111] (Fig. 3). While increased seizure activity was consistent with the increased kainate-induced neuronal death in the hippocampus of *bak*$^{-/-}$ mice, why would *bak*-deficiency cause a seizure phenotype within 5 minutes of treatment, days before neuronal death would be detectable? This seizure phenotype was also unexpected for a cell death regulator because other cell death regulators had been reported to alter neuronal survival in the same kainate seizure model but without any differences in seizure activity.[117,118] The implication was that the brains of *bak*$^{-/-}$ mice were somehow wired differently from normal mice. Electrophysiological studies confirmed altered spontaneous and evoked activities in acute brain slices prepared from wild type and *bak*$^{-/-}$ mice.[111] At the same time using an alternate approach, recombinant Bcl-x$_L$ was found to dramatically alter neuronal activity in the squid giant synapse (see below), suggesting that BCL-2 family proteins could have a profound and direct effect on synaptic activity.[119,120]

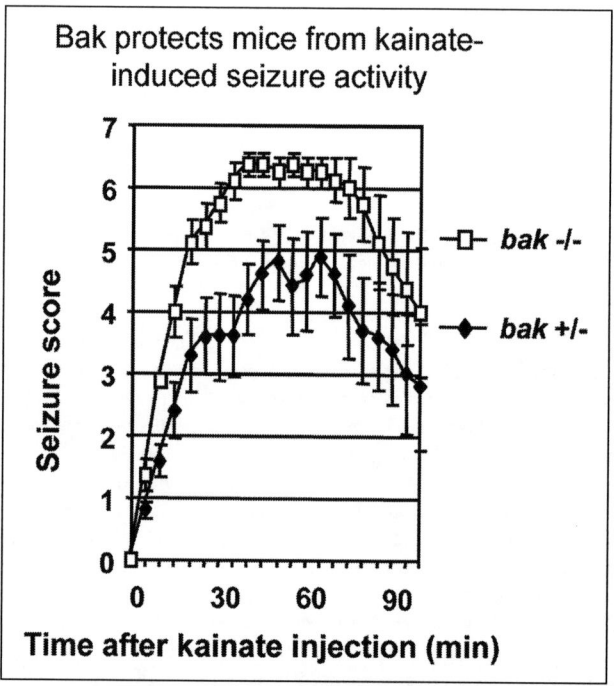

Figure 3. Bak protects against kainate-induced seizure activity. Kainate was injected intraperitoneally into Bak knockout and heterozygous littermates and seizure activity was scored as previously reported.[111] Reprinted from Fannjiang Yet al. Dev Cell 2003; 4:575-585, with permission from Elsevier.

Taken together, these unexpected findings triggered a reassessment of BCL-2 family functions in neurons that are independent of their roles in apoptosis. The other exciting aspect of these discoveries was the implication that BCL-2 proteins could be of considerable importance in many neurological disorders, where the decline in synaptic function and the onset of patient symptoms long precedes neuronal cell death. Thus, BCL-2 family proteins potentially influence the initial steps in neurodegeneration[65,111,119,121] and their role in neuronal dysfunction can no longer be so easily dismissed.

BCL-2 Family Proteins Regulate Synaptic Activity

The current working model is that BCL-2 family members act on mitochondria to facilitate synaptic changes in neurons. A startling but telling early finding was that recombinant Bcl-x$_L$ protein purified from *E. coli* and injected into the presynaptic cell at the squid giant synapse could rescue a lazy response by the neuron to a 0.03 Hz stimulus within 5-15 min after injection.[119] In this model system, Bcl-x$_L$ could also speed the recovery of neurotransmission following an intense repetitive stimulus that induces rapid neuronal firing (tetanus). Furthermore, run-down of synaptic activity induced by hypoxia could be significantly delayed by Bcl-x$_L$ and could be mimicked by an N-terminally truncated Bcl-x$_L$ protein corresponding to its pro-death caspase cleavage fragment.[120,122] Interestingly, recombinant Bax elicited either of two types of electrophysiological responses in squid, one similar to full-length Bcl-x$_L$, the other similar to truncated Bcl-x$_L$.[123] This dichotomy is reminiscent of the pre-activated and activated forms of Bax during cell death.

Because the effects of injected recombinant Bcl-x$_L$ in the squid giant synapse can be mimicked and occluded by injection of ATP, this raised the possibility that the effects of Bcl-x$_L$ on mitochondrial energetics was responsible for stimulating and preserving synaptic activity.[119] This idea was consistent with other studies suggesting that Bcl-x$_L$ can sustain VDAC-mediated translocation of ATP from the intermembrane space into the cytosol following a death stimulus.[124-126] If we overlook the long-held assumption that ATP is in abundance inside cells (but is perhaps more like money in an economy), then ATP could be a key rate-limiting component of synaptic transmission. Therefore, an increased supply of ATP induced by Bcl-x$_L$ could facilitate the energy-requiring processes of neurotransmitter production, packaging, release and vesicle recycling at the synapse. A critical function of Bcl-x$_L$ must occur within minutes to explain its effects in the squid model. Distinguishing direct from indirect effects of Bcl-x$_L$ will also be challenging and the underlying mechanisms are not yet known.

Mitochondrial Dynamics in Neuronal Function

Longer-term effects of Bcl-x$_L$ are also likely to significantly impact on neuronal activity, as evident in cultured neurons overexpressing Bcl-x$_L$ for days. These include changes in shape and localization of mitochondria (see below), which seem unlikely to explain the rapid effects of Bcl-x$_L$ in the squid model discussed above. However, both the immediate and long-term consequences of Bcl-x$_L$ in neurons could be reflections of a unifying and underlying biochemical activity. These underlying actions of Bcl-x$_L$ on mitochondria could also contribute importantly to its anti-death activity following a death stimulus if Bcl-x$_L$ and other anti-death family members allow the cell to meet new energy requirements following damage.

There is mounting evidence that BCL-2 family proteins regulate mitochondrial shape changes and mitochondrial localization in healthy cells.[127-129] Bcl-x$_L$ and its *C. elegans* counterpart, CED-9, both localize to mitochondria and are implicated in facilitating mitochondrial fusion and fission.[130] This is based in part on their ability to genetically interact and/or associate in (coprecipitated) protein complexes with the large dynamin-like GTPases Drp1 and Mfn2 that can localize to outer mitochondrial membranes and mediate organelle fission and fusion, respectively.[37,130-132] Although there is evidence for direct protein-protein interactions between these GTPases and BCL-2 family proteins, this point is not universally supported.[133,134]

Taking advantage of the long, thin processes of cultured neurons, the effects of Bcl-x$_L$ on mitochondrial fission and fusion were evaluated by directly counting these events using a fluorescence microscopy approach (DROF) combined with computational analyses.[65,135] The results are in general agreement with the observations of other investigators,[136-138] but assigning values to the rates of mitochondrial fission and fusion revealed unexpected findings. Bcl-x$_L$ increased the rate of fusion, consistent with longer mitochondrial organelles, but also increased the rate of fission. Importantly, the rate of fission with or without Bcl-x$_L$ was higher than the rate of fusion, implying that mitochondria must grow in length/size to compensate for a fission:fusion ratio greater than 1.[65] Taking all measured and computed parameters into consideration, the results suggest that Bcl-x$_L$ increases mitochondrial biomass at steady state. This is consistent with a previous report that BCL-2 overexpression increases the matrix volume and complexity of mitochondria in neural cell lines.[139]

An alternative mechanism to explain the increase in mitochondrial biomass is that Bcl-x$_L$ inhibits mitochondrial degradation, presumably by inhibiting macro autophagy. This is consistent with the ability of Bcl-x$_L$ to interact with and inhibit Beclin, homologue of yeast ATG6 and mediator of an early step in the autophagy pathway.[36,140] Regardless of whether Bcl-x$_L$ increases biomass by increasing biogenesis and/or decreasing mitophagy, either mechanism must be compensated to achieve steady state conditions. That is, any Bcl-x$_L$-induced increase in mitochondrial biogenesis would have to be balanced by increased mitochondrial turnover, and Bcl-x$_L$-induced decrease in turnover would in turn have to compensated by suppressing basal biogenesis at steady state. The integration of mitochondrial fission and fusion rates with the rates of mitochondrial biogenesis and degradation are not yet understood but likely to be influenced by BCL-2 family proteins.

Conclusion

Mounting evidence suggests that BCL-2 family members engage in critical functions in addition to regulating apoptosis, including the regulation of synaptic activity in healthy neurons. These noncanonical roles evoke the hypothesis that loss of BCL-2 family member function may directly contribute to neuronal dysfunction prior to the initiation of any cell death program. The connection is not known between cell death modifier functions and the noncanonical roles of BCL-2 family proteins that affect normal neuronal physiology. Small molecules that bind in the cleft of Bcl-xL and inhibit its anti-death activity may also inhibit these noncanonical functions. Unlike ABT-737, which does not cross the blood brain barrier, other small molecular inhibitors of BCL-2 proteins have neurological side effects, though off-site targets are possible. Given the evidence that Bcl-xL regulates mitochondrial energetics, dynamics, and increases localization of mitochondria to synapses, it is reasonable to speculate that inhibition of Bcl-xL in the brain may lead to alterations in mitochondrial energetics and/or morphology, which may underlie synaptic abnormalities. While experimental evidence points toward noncanonical functions for many BCL-2 family members, further characterization of these novel mechanisms and associated functions will yield important insight into the role of BCL-2 family members in health and disease.

References

1. Tsujimoto Y, Cossman J, Jaffe E et al. Involvement of the bcl-2 gene in human follicular lymphoma. Science 1985; 228:1440-1443.
2. Ohno H, Fukuhara S, Takahashi R et al. C-yes and bcl-2 genes located on 18q21.3 in a follicular lymphoma cell line carrying a t(14;18) chromosomal translocation. Int J Cancer 1987; 39:785-788.
3. Graninger WB, Seto M, Boutain B et al. Expression of bcl-2 and bcl-2-ig fusion transcripts in normal and neoplastic cells. J Clin Invest 1987; 80:1512-1515.
4. Weiss LM, Warnke RA, Sklar J et al. Molecular analysis of the t(14;18) chromosomal translocation in malignant lymphomas. N Engl J Med 1987; 317:1185-1189.
5. Hengartner MO, Ellis RE, Horvitz HR. Caenorhabditis elegans gene ced-9 protects cells from programmed cell death. Nature 1992; 356:494-499.
6. Yuan J, Shaham S, Ledoux S et al. The c. Elegans cell death gene ced-3 encodes a protein similar to mammalian interleukin-1 beta-converting enzyme. Cell 1993; 75:641-652.
7. Vaux DL, Cory S, Adams JM. BCL-2 gene promotes haemopoietic cell survival and cooperates with c-myc to immortalize preb cells. Nature 1988; 335:440-442.
8. Henderson S, Huen D, Rowe M et al. Epstein-barr virus-coded bhrf1 protein, a viral homologue of bcl-2, protects human b-cells from programmed cell death. Proc Natl Acad Sci USA 1993; 90:8479-8483.

9. Adams JM, Cory S. The bcl-2 protein family: Arbiters of cell survival. Science 1998; 281:1322-1326.

10. Youle RJ, Strasser A. The bcl-2 protein family: Opposing activities that mediate cell death. Nat Rev Mol Cell Biol 2008; 9:47-59.

11. Fletcher JI, Meusburger S, Hawkins CJ et al. Apoptosis is triggered when prosurvival bcl-2 proteins cannot restrain bax. Proc Natl Acad Sci USA 2008; 105:18081-18087.

12. Kuwana T, Mackey MR, Perkins G et al. Bid, bax and lipids cooperate to form supramolecular openings in the outer mitochondrial membrane. Cell 2002; 111:331-342.

13. Antonsson B, Montessuit S, Lauper S et al. Bax oligomerization is required for channel-forming activity in liposomes and to trigger cytochrome c release from mitochondria. Biochem J 2000; 345(Pt 2):271-278.

14. Budihardjo I, Oliver H, Lutter M et al. Biochemical pathways of caspase activation during apoptosis. Annu Rev Cell Dev Biol 1999; 15:269-290.

15. Antonsson B, Conti F, Ciavatta A et al. Inhibition of bax channel-forming activity by bcl-2. Science 1997; 277:370-372.

16. Kvansakul M, Yang H, Fairlie WD et al. Vaccinia virus anti-apoptotic f11 is a novel bcl-2-like domain-swapped dimer that binds a highly selective subset of bh3-containing death ligands. Cell Death Differ 2008; 15:1564-1571.

17. Graham SC, Bahar MW, Cooray S et al. Vaccinia virus proteins a52 and b14 share a bcl-2-like fold but have evolved to inhibit nf-kappab rather than apoptosis. PLoS Pathog 2008; 4:e1000128.

18. Galindo KA, Lu WJ, Park JH et al. The bax/bak ortholog in drosophila, debcl, exerts limited control over programmed cell death. Development 2009; 136:275-283.

19. Bellows DS, Howell M, Pearson C et al. Epstein-barr virus balf1 is a bcl-2-like antagonist of the herpesvirus antiapoptotic bcl-2 proteins. J Virol 2002; 76:2469-2479.

20. Huang DC, Strasser A. Bh3-only proteins-essential initiators of apoptotic cell death. Cell 2000; 103:839-842.

21. Cheng EH, Wei MC, Weiler S et al. BCL-2, bcl-x(l) sequester bh3 domain-only molecules preventing bax- and bak-mediated mitochondrial apoptosis. Mol Cell 2001; 8:705-711.

22. Letai A, Bassik MC, Walensky LD et al. Distinct bh3 domains either sensitize or activate mitochondrial apoptosis, serving as prototype cancer therapeutics. Cancer Cell 2002; 2:183-192.

23. Billen LP, Kokoski CL, Lovell JF et al. Bcl-xl inhibits membrane permeabilization by competing with bax. PLoS Biol 2008; 6:e147.

24. Day CL, Smits C, Fan FC et al. Structure of the bh3 domains from the p53-inducible bh3-only proteins noxa and puma in complex with mcl-1. J Mol Biol 2008; 380:958-971.

25. Hinds MG, Smits C, Fredericks-Short R et al. Bim, bad and bmf: Intrinsically unstructured bh3-only proteins that undergo a localized conformational change upon binding to prosurvival bcl-2 targets. Cell Death Differ 2007; 14:128-136.

26. Oltersdorf T, Elmore SW, Shoemaker AR et al. An inhibitor of bcl-2 family proteins induces regression of solid tumours. Nature 2005; 435:677-681.

27. Deng J, Carlson N, Takeyama K et al. Bh3 profiling identifies three distinct classes of apoptotic blocks to predict response to abt-737 and conventional chemotherapeutic agents. Cancer Cell 2007; 12:171-185.

28. Park CM, Bruncko M, Adickes J et al. Discovery of an orally bioavailable small molecule inhibitor of prosurvival b-cell lymphoma 2 proteins. J Med Chem 2008; 51:6902-6915.

29. Danial NN, Walensky LD, Zhang CY et al. Dual role of proapoptotic bad in insulin secretion and beta cell survival. Nat Med 2008; 14:144-153.

30. Gavathiotis E, Suzuki M, Davis ML et al. Bax activation is initiated at a novel interaction site. Nature 2008; 455:1076-1081.

31. Zhong Q, Gao W, Du F et al. Mule/arf-bp1, a bh3-only e3 ubiquitin ligase, catalyzes the polyubiquitination of mcl-1 and regulates apoptosis. Cell 2005; 121:1085-1095.

32. Oberstein A, Jeffrey PD, Shi Y. Crystal structure of the bcl-xl-beclin 1 peptide complex: Beclin 1 is a novel bh3-only protein. J Biol Chem 2007; 282:13123-13132.

33. Mihara M, Erster S, Zaika A et al. P53 has a direct apoptogenic role at the mitochondria. Mol Cell 2003; 11:577-590.

34. White C, Li C, Yang J et al. The endoplasmic reticulum gateway to apoptosis by bcl-x(l) modulation of the insp3r. Nat Cell Biol 2005; 7:1021-1028.

35. Puthalakath H, Villunger A, O'Reilly LA et al. Bmf: A proapoptotic bh3-only protein regulated by interaction with the myosin v actin motor complex, activated by anoikis. Science 2001; 293:1829-1832.

36. Pattingre S, Tassa A, Qu X et al. BCL-2 antiapoptotic proteins inhibit beclin 1-dependent autophagy. Cell 2005; 122:927-939.

37. Delivani P, Adrain C, Taylor RC et al. Role for ced-9 and egl-1 as regulators of mitochondrial fission and fusion dynamics. Mol Cell 2006; 21:761-773.

38. Guo JY, Yamada A, Kajino T et al. Aven-dependent activation of atm following DNA damage. Curr Biol 2008; 18:933-942.

39. Cheng EH, Levine B, Boise LH et al. Bax-independent inhibition of apoptosis by bcl-xl. Nature 1996; 379:554-556.
40. Cheng EH, Sheiko TV, Fisher JK et al. Vdac2 inhibits bak activation and mitochondrial apoptosis. Science 2003; 301:513-517.
41. Kim H, Rafiuddin-Shah M, Tu HC et al. Hierarchical regulation of mitochondrion-dependent apoptosis by bcl-2 subfamilies. Nat Cell Biol 2006; 8:1348-1358.
42. Hsu YT, Youle RJ. Nonionic detergents induce dimerization among members of the bcl-2 family. J Biol Chem 1997; 272:13829-13834.
43. Basanez G, Nechushtan A, Drozhinin O et al. Bax, but not bcl-xl, decreases the lifetime of planar phospholipid bilayer membranes at subnanomolar concentrations. Proc Natl Acad Sci USA 1999; 96:5492-5497.
44. Basanez G, Sharpe JC, Galanis J et al. Bax-type apoptotic proteins porate pure lipid bilayers through a mechanism sensitive to intrinsic monolayer curvature. J Biol Chem 2002; 277:49360-49365.
45. Kane DJ, Ord T, Anton R et al. Expression of bcl-2 inhibits necrotic neural cell death. J Neurosci Res 1995; 40:269-275.
46. Dickman MB, Park YK, Oltersdorf T et al. Abrogation of disease development in plants expressing animal antiapoptotic genes. Proc Natl Acad Sci USA 2001; 98:6957-6962.
47. Krajewska M, Mai JK, Zapata JM et al. Dynamics of expression of apoptosis-regulatory proteins bid, bcl-2, bcl-x, bax and bak during development of murine nervous system. Cell Death Differ 2002; 9:145-157.
48. Krajewski S, Krajewska M, Shabaik A et al. Immunohistochemical determination of in vivo distribution of bax, a dominant inhibitor of bcl-2. Am J Pathol 1994; 145:1323-1336.
49. Krajewski S, Krajewska M, Shabaik A et al. Immunohistochemical analysis of in vivo patterns of bcl-x expression. Cancer Res 1994; 54:5501-5507.
50. Arbour N, Vanderluit JL, Le Grand JN et al. Mcl-1 is a key regulator of apoptosis during cns development and after DNA damage. J Neurosci 2008; 28:6068-6078.
51. O'Reilly LA, Print C, Hausmann G et al. Tissue expression and subcellular localization of the pro-survival molecule bcl-w. Cell Death Differ 2001; 8:486-494.
52. Uo T, Kinoshita Y, Morrison RS. Neurons exclusively express n-bak, a bh3 domain-only bak isoform that promotes neuronal apoptosis. J Biol Chem 2005; 280:9065-9073.
53. Fu NY, Sukumaran SK, Kerk SY et al. Baxbeta: A constitutively active human bax isoform that is under tight regulatory control by the proteasomal degradation mechanism. Mol Cell 2009; 33:15-29.
54. Lindsten T, Ross AJ, King A et al. The combined functions of proapoptotic bcl-2 family members bak and bax are essential for normal development of multiple tissues. Mol Cell 2000; 6:1389-1399.
55. Whitmore AV, Lindsten T, Raff MC et al. The proapoptotic proteins bax and bak are not involved in wallerian degeneration. Cell Death Differ 2003; 10:260-261.
56. Glebova NO, Ginty DD. Heterogeneous requirement of ngf for sympathetic target innervation in vivo. J Neurosci 2004; 24:743-751.
57. Liu QA, Shio H. Mitochondrial morphogenesis, dendrite development and synapse formation in cerebellum require both bcl-w and the glutamate receptor delta2. PLoS Genet 2008; 4:e1000097.
58. Murphy B, Dunleavy M, Shinoda S et al. BCL-2 protects hippocampus during experimental status epilepticus. Am J Pathol 2007; 171:1258-1268.
59. Veis DJ, Sorenson CM, Shutter JR et al. BCL-2-deficient mice demonstrate fulminant lymphoid apoptosis, polycystic kidneys and hypopigmented hair. Cell 1993; 75:229-240.
60. Merry DE, Veis DJ, Hickey WF et al. BCL-2 protein expression is widespread in the developing nervous system and retained in the adult pns. Development 1994; 120:301-311.
61. Michaelidis TM, Sendtner M, Cooper JD et al. Inactivation of bcl-2 results in progressive degeneration of motoneurons, sympathetic and sensory neurons during early postnatal development. Neuron 1996; 17:75-89.
62. Motoyama N, Wang F, Roth KA et al. Massive cell death of immature hematopoietic cells and neurons in bcl-x-deficient mice. Science 1995; 267:1506-1510.
63. Zhang J, Chen YB, Hardwick JM et al. Magnetic resonance diffusion tensor microimaging reveals a role for bcl-x in brain development and homeostasis. J Neurosci 2005; 25:1881-1888.
64. Savitt JM, Jang SS, Mu W et al. Bcl-x is required for proper development of the mouse substantia nigra. J Neurosci 2005; 25:6721-6728.
65. Berman SB, Chen YB, Qi B et al. Bcl-xl increases mitochondrial fission, fusion and biomass in neurons. J Cell Biol 2009; 184:707-719.
66. Shindler KS, Latham CB, Roth KA. Bax deficiency prevents the increased cell death of immature neurons in bcl-x-deficient mice. J Neurosci 1997; 17:3112-3119.
67. Akhtar RS, Ness JM, Roth KA. BCL-2 family regulation of neuronal development and neurodegeneration. Biochim Biophys Acta 2004; 1644:189-203.
68. Roth KA, D'Sa C. Apoptosis and brain development. Ment Retard Dev Disabil Res Rev 2001; 7:261-266.

69. Akhtar RS, Geng Y, Klocke BJ et al. Bh3-only proapoptotic bcl-2 family members noxa and puma mediate neural precursor cell death. J Neurosci 2006; 26:7257-7264.
70. Geng Y, Akhtar RS, Shacka JJ et al. P53 transcription-dependent and -independent regulation of cerebellar neural precursor cell apoptosis. J Neuropathol Exp Neurol 2007; 66:66-74.
71. Akhtar RS, Klocke BJ, Strasser A et al. Loss of bh3-only protein bim inhibits apoptosis of hemopoietic cells in the fetal liver and male germ cells but not neuronal cells in bcl-x-deficient mice. J Histochem Cytochem 2008; 56:921-927.
72. Wagner KU, Claudio E, Rucker EB, 3rd et al. Conditional deletion of the bcl-x gene from erythroid cells results in hemolytic anemia and profound splenomegaly. Development 2000; 127:4949-4958.
73. Qi B, Hardwick JM. A bcl-xl timer sets platelet life span. Cell 2007; 128:1035-1036.
74. Mason KD, Carpinelli MR, Fletcher JI et al. Programmed anuclear cell death delimits platelet life span. Cell 2007; 128:1173-1186.
75. Doonan F, Donovan M, Gomez-Vicente V et al. Bim expression indicates the pathway to retinal cell death in development and degeneration. J Neurosci 2007; 27:10887-10894.
76. Hetz C, Thielen P, Fisher J et al. The proapoptotic bcl-2 family member bim mediates motoneuron loss in a model of amyotrophic lateral sclerosis. Cell Death Differ 2007; 14:1386-1389.
77. McKernan DP, Cotter TG. A critical role for bim in retinal ganglion cell death. J Neurochem 2007; 102:922-930.
78. Bouillet P, Cory S, Zhang LC et al. Degenerative disorders caused by bcl-2 deficiency prevented by loss of its bh3-only antagonist bim. Dev Cell 2001; 1:645-653.
79. Lewis J, Wesselingh SL, Griffin DE et al. Alphavirus-induced apoptosis in mouse brains correlates with neurovirulence. J Virol 1996; 70:1828-1835.
80. Lewis J, Oyler GA, Ueno K et al. Inhibition of virus-induced neuronal apoptosis by bax. Nat Med 1999; 5:832-835.
81. Levine B, Huang Q, Isaacs JT et al. Conversion of lytic to persistent alphavirus infection by the bcl-2 cellular oncogene. Nature 1993; 361:739-742.
82. Levine B, Goldman JE, Jiang HH et al. Bcl-2 protects mice against fatal alphavirus encephalitis. Proc Natl Acad Sci USA 1996; 93:4810-4815.
83. Irusta PM, Hardwick JM. Neuronal apoptosis pathways in sindbis virus encephalitis. Prog Mol Subcell Biol 2004; 36:71-93.
84. Cheng EH, Kirsch DG, Clem RJ et al. Conversion of bcl-2 to a bax-like death effector by caspases. Science 1997; 278:1966-1968.
85. Kirsch DG, Doseff A, Chau BN et al. Caspase-3-dependent cleavage of bcl-2 promotes release of cyto-chrome c. J Biol Chem 1999; 274:21155-21161.
86. Clem RJ, Cheng EH, Karp CL et al. Modulation of cell death by bcl-xl through caspase interaction. Proc Natl Acad Sci USA 1998; 95:554-559.
87. Grandgirard D, Studer E, Monney L et al. Alphaviruses induce apoptosis in bcl-2-overexpressing cells: Evidence for a caspase-mediated, proteolytic inactivation of bcl-2. EMBO J 1998; 17:1268-1278.
88. Basanez G, Zhang J, Chau BN et al. Pro-apoptotic cleavage products of bcl-xl form cytochrome c-conducting pores in pure lipid membranes. J Biol Chem 2001; 276:31083-31091.
89. Kelekar A, Thompson CB. BCL-2-family proteins: The role of the bh3 domain in apoptosis. Trends Cell Biol 1998; 8:324-330.
90. Seo SY, Chen YB, Ivanovska I et al. Bad is a pro-survival factor prior to activation of its pro-apoptotic function. J Biol Chem 2004; 279:42240-42249.
91. Condorelli F, Salomoni P, Cotteret S et al. Caspase cleavage enhances the apoptosis-inducing effects of bad. Mol Cell Biol 2001; 21:3025-3036.
92. Datta SR, Dudek H, Tao X et al. Akt phosphorylation of bad couples survival signals to the cell-intrinsic death machinery. Cell 1997; 91:231-241.
93. Wood DE, Thomas A, Devi LA et al. Bax cleavage is mediated by calpain during drug-induced apoptosis. Oncogene 1998; 17:1069-1078.
94. Chen D, Zhou Q. Caspase cleavage of bimel triggers a positive feedback amplification of apoptotic signaling. Proc Natl Acad Sci USA 2004; 101:1235-1240.
95. Gomez-Bougie P, Wuilleme-Toumi S, Menoret E et al. Noxa up-regulation and mcl-1 cleavage are associated to apoptosis induction by bortezomib in multiple myeloma. Cancer Res 2007; 67:5418-5424.
96. Michels J, Johnson PW, Packham G. Mcl-1. Int J Biochem Cell Biol 2005; 37:267-271.
97. Michels J, O'Neill JW, Dallman CL et al. Mcl-1 is required for akata6 b-lymphoma cell survival and is converted to a cell death molecule by efficient caspase-mediated cleavage. Oncogene 2004; 23:4818-4827.
98. Li H, Zhu H, Xu CJ et al. Cleavage of bid by caspase 8 mediates the mitochondrial damage in the fas pathway of apoptosis. Cell 1998; 94:491-501.
99. Gil-Parrado S, Fernandez-Montalvan A, Assfalg-Machleidt I et al. Ionomycin-activated calpain triggers apoptosis. A probable role for bcl-2 family members. J Biol Chem 2002; 277:27217-27226.

100. Qi B, Hardwick JM. BCL-2 turns deadly. Nat Chem Biol 2008; 4:722-723.
101. Kolluri SK, Zhu X, Zhou X et al. A short nur77-derived peptide converts bcl-2 from a protector to a killer. Cancer Cell 2008; 14:285-298.
102. Lin B, Kolluri SK, Lin F et al. Conversion of bcl-2 from protector to killer by interaction with nuclear orphan receptor nur77/tr3. Cell 2004; 116:527-540.
103. Luciano F, Krajewska M, Ortiz-Rubio P et al. Nur77 converts phenotype of bcl-b, an antiapoptotic protein expressed in plasma cells and myeloma. Blood 2007; 109:3849-3855.
104. Thompson J, Winoto A. During negative selection, nur77 family proteins translocate to mitochondria where they associate with bcl-2 and expose its proapoptotic bh3 domain. J Exp Med 2008; 205:1029-1036.
105. Hickman JA, Hardwick JM, Kaczmarek LK et al. Bcl-xl inhibitor abt-737 reveals a dual role for bcl-xl in synaptic transmission. J Neurophysiol 2008; 99:1515-1522.
106. Ubol S, Tucker PC, Griffin DE et al. Neurovirulent strains of alphavirus induce apoptosis in bcl-2-expressing cells: Role of a single amino acid change in the e2 glycoprotein. Proc Natl Acad Sci USA 1994; 91:5202-5206.
107. Levine B, Hardwick JM, Griffin DE. Persistence of alphaviruses in vertebrate hosts. Trends Microbiol 1994; 2:25-28.
108. Griffin DE, Hardwick JM. Perspective: Virus infections and the death of neurons. Trends Microbiol 1999; 7:155-160.
109. Griffin DE, Levine B, Ubol S et al. The effects of alphavirus infection on neurons. Ann Neurol 1994; 35 Suppl:S23-27.
110. Hardwick JM, Levine B. Sindbis virus vector system for functional analysis of apoptosis regulators. Methods Enzymol 2000; 322:492-508.
111. Fannjiang Y, Kim CH, Huganir RL et al. Bak alters neuronal excitability and can switch from anti- to pro-death function during postnatal development. Dev Cell 2003; 4:575-585.
112. Middleton G, Davies AM. Populations of ngf-dependent neurones differ in their requirement for bax to undergo apoptosis in the absence of ngf/trka signalling in vivo. Development 2001; 128:4715-4728.
113. Middleton G, Nunez G, Davies AM. Bax promotes neuronal survival and antagonises the survival effects of neurotrophic factors. Development 1996; 122:695-701.
114. Senoo-Matsuda N, Igaki T, Miura M. Bax-like protein drob-1 protects neurons from expanded polyglutamine-induced toxicity in drosophila. EMBO J 2005; 24:2700-2713.
115. Chew SK, Chen P, Link N et al. Genome-wide silencing in drosophila captures conserved apoptotic effectors. Nature 2009.
116. Sevrioukov EA, Burr J, Huang EW et al. Drosophila bcl-2 proteins participate in stress-induced apoptosis, but are not required for normal development. Genesis 2007; 45:184-193.
117. Morrison RS, Wenzel HJ, Kinoshita Y et al. Loss of the p53 tumor suppressor gene protects neurons from kainate-induced cell death. J Neurosci 1996; 16:1337-1345.
118. Holcik M, Thompson CS, Yaraghi Z et al. The hippocampal neurons of neuronal apoptosis inhibitory protein 1 (naip1)-deleted mice display increased vulnerability to kainic acid-induced injury. Proc Natl Acad Sci USA 2000; 97:2286-2290.
119. Jonas EA, Hoit D, Hickman JA et al. Modulation of synaptic transmission by the bcl-2 family protein bcl-xl. J Neurosci 2003; 23:8423-8431.
120. Jonas EA, Hickman JA, Chachar M et al. Proapoptotic n-truncated bcl-xl protein activates endogenous mitochondrial channels in living synaptic terminals. Proc Natl Acad Sci USA 2004; 101:13590-13595.
121. Karbowski M, Norris KL, Cleland MM et al. Role of bax and bak in mitochondrial morphogenesis. Nature 2006; 443:658-662.
122. Jonas EA, Hickman JA, Hardwick JM et al. Exposure to hypoxia rapidly induces mitochondrial channel activity within a living synapse. J Biol Chem 2005; 280:4491-4497.
123. Jonas EA, Hardwick JM, Kaczmarek LK. Actions of bax on mitochondrial channel activity and on synaptic transmission. Antioxid Redox Signal 2005; 7:1092-1100.
124. Vander Heiden MG, Thompson CB. BCL-2 proteins: Regulators of apoptosis or of mitochondrial homeostasis? Nat Cell Biol 1999; 1:E209-216.
125. Vander Heiden MG, Li XX, Gottleib E et al. Bcl-xl promotes the open configuration of the voltage-dependent anion channel and metabolite passage through the outer mitochondrial membrane. J Biol Chem 2001; 276:19414-19419.
126. Vander Heiden MG, Chandel NS, Williamson EK et al. Bcl-xl regulates the membrane potential and volume homeostasis of mitochondria. Cell 1997; 91:627-637.
127. Frank S. Dysregulation of mitochondrial fusion and fission: An emerging concept in neurodegeneration. Acta Neuropathol 2006; 111:93-100.
128. Detmer SA, Chan DC. Functions and dysfunctions of mitochondrial dynamics. Nat Rev Mol Cell Biol 2007; 8:870-879.

129. Knott AB, Perkins G, Schwarzenbacher R et al. Mitochondrial fragmentation in neurodegeneration. Nat Rev Neurosci 2008; 9:505-518.
130. Jagasia R, Grote P, Westermann B et al. Drp-1-mediated mitochondrial fragmentation during egl-1-induced cell death in c. Elegans. Nature 2005; 433:754-760.
131. Li H, Chen Y, Jones AF et al. Bcl-xl induces drp1-dependent synapse formation in cultured hippocampal neurons. Proc Natl Acad Sci USA 2008; 105:2169-2174.
132. Tan FJ, Husain M, Manlandro CM et al. Ced-9 and mitochondrial homeostasis in c. Elegans muscle. J Cell Sci 2008; 121:3373-3382.
133. Parone PA, James DI, Da Cruz S et al. Inhibiting the mitochondrial fission machinery does not prevent bax/bak-dependent apoptosis. Mol Cell Biol 2006; 26:7397-7408.
134. Breckenridge DG, Kang BH, Xue D. BCL-2 proteins egl-1 and ced-9 do not regulate mitochondrial fission or fusion in caenorhabditis elegans. Curr Biol 2009.
135. Berman SB, Pineda FJ, Hardwick JM. Mitochondrial fission and fusion dynamics: The long and short of it. Cell Death Differ 2008; 15:1147-1152.
136. Karbowski M, Youle RJ. Dynamics of mitochondrial morphology in healthy cells and during apoptosis. Cell Death Differ 2003; 10:870-880.
137. Sheridan C, Delivani P, Cullen SP et al. Bax- or bak-induced mitochondrial fission can be uncoupled from cytochrome c release. Mol Cell 2008; 31:570-585.
138. Twig G, Elorza A, Molina AJ et al. Fission and selective fusion govern segregation and elimination by autophagy. EMBO J 2008; 27:433-446.
139. Kowaltowski AJ, Cosso RG, Campos CB et al. Effect of bcl-2 overexpression on mitochondrial structure and function. J Biol Chem 2002; 277:42802-42807.
140. Maiuri MC, Criollo A, Tasdemir E et al. Bh3-only proteins and bh3 mimetics induce autophagy by competitively disrupting the interaction between beclin 1 and bcl-2/bcl-x(l). Autophagy 2007; 3:374-376.

CHAPTER 8

Mitochondria on Guard:
Role of Mitochondrial Fusion and Fission in the Regulation of Apoptosis

Mariusz Karbowski

Abstract

Mitochondria are highly dynamic organelles that constantly change shape and structure in response to different stimuli and metabolic demands of the cell. Mitochondrial structure in the cell is predominantly regulated by cycles of fusion and fission. These two processes are tightly regulated and under physiological conditions, mitochondrial fusion is evenly counterbalanced by fission. During apoptosis, mitochondria undergo extensive fragmentation, which precedes caspase activation, whereas inhibition of the mitochondrial fission machinery blocks or delays cell death. Aberrant mitochondrial fusion and fission have also emerged as important mechanisms in the development of disease.

In this chapter we will briefly discuss mechanisms of mitochondrial fusion and fission in mammalian cells, the mitochondrial morphogenesis proteins that may be involved in these processes and role of these processes in cell viability. This will be followed by a review of recent work connecting mitochondrial morphogenesis proteins with the progression of the mitochondrial steps in apoptosis, as well as stimulating findings showing that certain proteins associated with apoptosis regulation can also participate in the regulation of mitochondrial fusion and fission in healthy cells.

Introduction

Mitochondrial architecture in the cell alternates between long tubules and small round structures, resulting in a dynamic organizational equilibrium. This equilibrium is achieved through the continuously occurring but opposing processes of mitochondrial fusion and fission. Mitochondrial fusion and fission are tightly regulated and under physiological conditions mitochondrial fusion is evenly counterbalanced by fission. Aberrant mitochondrial fusion and fission have recently emerged as important mechanisms in the development of disease. The wide spectrum of mitochondrial diseases that typically involved deficiencies of the oxidative phosphorylation, now includes genetic and biochemical alterations of mitochondrial fusion and fission. Furthermore, the altered activities of mitochondrial morphogenesis-associated proteins have also been implicated in the regulation of the mitochondrial steps in apoptosis and the stress response of the cell.

Regulation of Mitochondrial Fusion and Fission

Mitochondrial Fusion

Mitochondrial fusion results in the enlargement of mitochondria by tethering and joining together two adjacent mitochondria. One unique feature of this process is the necessity of syn-

*Mariusz Karbowski—University of Maryland Biotechnology Institute, Medical Biotechnology Center, 725 W. Lombard St, Baltimore, Maryland 21201, USA.
Email: karbowsk@umbi.umd.edu

BCL-2 Protein Family: Essential Regulators of Cell Death, edited by Claudio Hetz.
©2010 Landes Bioscience and Springer Science+Business Media.

chronizing integration of two double bilayer systems from the fusing mitochondria. While, the fusion events of the OMM and the IMM are usually synchronized, under certain conditions the OMM can fuse without subsequent fusion of the IMM.[1] Dissipation of the mitochondrial membrane potential ($\Delta\Psi_m$) appears to selectively inhibit IMM fusion.[1] This, in addition to the identification of yeast mitochondrial fusion intermediates in vitro,[2] indicates that the OMM and IMM may fuse in successive and independent reactions. OMM fusion is regulated by Mitofusins (Mfn) 1 and 2, two large GTPases in the OMM.[3] Mfns are inserted in the OMM through two transmembrane domains and both the N- (GTPase domain) and the C- [heptad repeat regions (HR2)] terminal parts of these proteins are exposed to the cytosolic side of the OMM. The HR2s participate in the tethering of two fusing mitochondria through antiparallel coiled-coil hydrophobic interactions,[3-5] probably the initial step in mitochondrial fusion. Although each Mitofusin appears to possess specific and nonredundant roles,[4,6] the presence of both Mfn1 and Mfn2 is imperative for the maintenance of normal rates of mitochondrial fusion. Single knockouts of either Mfn1 or Mfn2 exhibit decreases in mitochondrial fusion rates[3,7] and mitochondrial fragmentation. However, the resulting mitochondrial shapes and sizes are Mfn1 or Mfn2 knockdown specific.[3] The depletion of Mfn1 induces small vesicular mitochondria broadly dispersed in the cell; whereas, the lack of Mfn2 results in larger mitochondrial vesicles concentrated around the nucleus.[3] Consequently, the role of each Mfn protein in mitochondrial fusion regulation might differ. Supporting this notion, it has been shown that Mfn1 is specifically required for GTP hydrolysis-dependent mitochondrial tethering, while Mfn2 is less efficient in this process and might act as a signaling GTPase.[4,6] IMM fusion is regulated by optic atrophy 1 (Opa1),[8] a dynamin related protein in the IMM. As in case of Mfns, knockdown of Opa1 leads to inhibition of mitochondrial fusion and mitochondrial fragmentation.[9] It has been proposed that an intermediate step in mitochondrial fusion might involve a pool of soluble Opa1 present in the intermembrane space,[10] which stabilizes transient Opa1 complexes. This mechanism, like Mfn-dependent mitochondrial fusion steps, is energy dependent, which is provided by the hydrolysis of GTP.[2] Merging of the two inner membranes is mediated by the physical interaction between IMM-localized Opa1. Mfn- and Opa1-dependent steps in mitochondrial fusion are likely synchronized by direct interactions between these proteins.[11]

In addition to Mfn1, Mfn2 and Opa1, other proteins have been implicated in the regulation of mitochondrial fusion in mammalian cells, including mitofusin binding protein (MiB),[12] mitochondria associated phospholipase D (mito-PLD)[13] and Stomatin-like protein 2 (Stoml2 also known as SLP2),[14,15] as well as mitochondrial proteases required for Opa1 processing, including the presinilin-associated rhomboid-like protein (PARL),[16] Yme1,[11,17,18] paraplegin[19] and the m-AAA protease complex.[20]

Mitochondrial Fission

Mitochondrial fission involves the constriction and cleavage of mitochondria by fission proteins, such as dynamin-related protein 1 (Drp1)[21] and Fis1.[22-25] The multimeric dynamin-like GTPase Drp1 harbors multiple motifs including GTP-binding, Middle and GTPase effector (GED) domains that are important for both intramolecular and intermolecular interactions. Drp1 knockdown or expression of dominant negative mutants of this protein [e.g., GTPAse domain mutant Drp1^{K38A}, or phosphomimetic mutant Drp1^{S637D}] lead to inhibition of mitochondrial fission and formation of abnormally elongated mitochondria, as well as highly interconnected mitochondrial networks.[21,26] Under normal growth conditions a major cellular fraction of Drp1 localizes to the cytosol and the translocation of Drp1 to the OMM is required to initiate mitochondrial fission.[21] Unlike most of the known OMM-associated proteins that are homogenously distributed along the OMM, the OMM associated pool of Drp1 forms discrete foci that often colocalize with progressive mitochondrial fission sites.[21] These structures are stabilized by SUMOylation[27] and probably destabilized by ubiquitination,[28] suggesting that these modifications of Drp1 could contribute to initiation and/or progression of the mitochondrial fission reaction. Furthermore, RNAi-mediated depletion of USP30, a mitochondria-localized ubiquitin isopeptidase, induced elongated and interconnected mitochondria, depending on the activities of Mfns, without changing the expression levels of the key regulators for mitochondrial dynamics.[29] Thus, deubiquitination of certain fission factor(s) is also implicated in the regulation or mitochondrial division. Phosphorylation of Drp1 is also critical for the mitochondrial activity of this protein. A quantity of kinases that phosphorylate Drp1, including cAMP-dependent protein kinase,[26,30] Ca^{2+}/calmodulin-dependent protein kinase I alpha (CaMKIalpha)[31] and cyclin-dependent kinase (Cdk1/cyclin B),[32] suggests a major role for phosphorylation in Drp1 and mitochondrial fission regulation.

Based on mitochondrial fission studies in *S. cerevisiae*, it has been proposed that in mammalian cells Fis1, an OMM anchored protein, acts as a mitochondrial receptor for Drp1.[23] Although the role of Fis1 in mitochondrial division is conserved,[23,24] the mechanism of Fis1 action in mammalian cells is not clear.

Recently, a number of proteins required for, or implicated in, mitochondrial division has been reported. These include: MTP18,[15] Mff (for mitochondrial fission factor),[33] as well as MARCH5[28,34,35] and MAPL,[36] a RING finger E3 ubiquitin ligases of the OMM and SenP5 a SUMO protease[37] and mitochondria-associated PTEN-induced kinase 1 (PINK1), linked to familial Parkinson's disease (PD).[38,39]

Mitochondrial Network Dynamics and Cell Homeostasis

Mutual Relations between Cell Homeostasis and Mitochondrial Network Dynamics

Studies on the cellular role of mitochondrial fusion and fission, as we discuss in this section, draw attention to the importance of mitochondrial morphogenesis and proteins regulating this process in cell and organism viability. Yet they also reveal how much remains to be learned about the mechanisms by which mitochondrial fusion and fission and specific mitochondrial morphogenesis proteins participate in the regulation of cellular homeostasis.

A number of mechanisms by which mitochondrial network dynamics can influence cell and mitochondrial function have been proposed. Mitochondrial fusion can facilitate intramitochondrial redistribution and complementation of mitochondrial DNA, as well as matrix contents and perhaps membrane proteins.[40] The cellular trafficking of the mitochondria also appears to rely on mitochondrial fusion and fission.[41] As a consequence, impairment in mitochondrial dynamics can induce abnormal accumulation of mitochondria in certain subcellular compartments, e.g., axons or perinuclear areas. Furthermore, elimination of dysfunctional mitochondria depends on local inhibition of mitochondrial fusion that enables the isolation of aberrant mitochondria from the bulk mitochondrial network, keeping them from re-entering the network. The isolated dysfunctional mitochondria are then targeted for degradation through the autophagy pathway.[42] In addition to fusion inhibition, activation of mitochondrial fission might also be needed for autophagic elimination of mitochondria as demonstrated in the *C. elegans* model, where the activity of Drp1 appears to facilitate mitochondrial elimination in dying cells.[43]

The relationship between cell physiology and mitochondrial network dynamics is extremely intricate and mutual. For example, declines in bioenergetic activity of mitochondria, including changes in mitochondrial membrane potential, concentration of ATP and ROS generation also govern mitochondrial shape.[44] On the other hand, alterations in the mitochondrial network, e.g., inhibition of mitochondrial fusion, result in defects in mitochondrial energy production.[45] As recently shown by Tondera et al,[46] when cells are exposed to selective stresses, including protein syntheis inhibition or UV irradiation, mitochondria can also undergo a process of extensive fusion and form a highly interconnected network. Stress-induced mitochondrial hyperfusion (SIMH) is independent of certain mitochondrial fusion proteins, including, Mfn2, Bax/Bak and prohibitins, but requires long form Opa1, Mfn1 and the mitochondrial inner membrane protein Slp-2.[46] Importantly, SIMH is accompanied by increased mitochondrial ATP production, suggesting that SIMH might represent a cell adaptation to metabolic perturbations and could be essential to allow overcoming transient, reversible metabolic insults. The molecular mechanism by which SIMH facilitates ATP generation is currently unknown.[46]

Mitochondrial Dynamics and Early Development

The regulation of mitochondrial fusion and fission has also had a significant impact on early development. The embryonic lethality induced by knockouts or mutations of fusion proteins Mfn1, Mfn2[3] and Opa1,[47] as well as Drp1, that is required for mitochondrial fission, indicate that the maintenance of a dynamic mitochondrial network is central for embryogenesis in mammals. Although Mfn1 and Mfn2 are ubiquitously expressed, the impact of embryonic defects in Mfn1$^{-/-}$ and Mfn2$^{-/-}$ differ.[3] In Mfn2$^{-/-}$ mice a disruption in placental development, most obviously in the paucity of trophoblast giant cells, has been reported.[3] Since similar defects were not obvious in Mfn1$^{-/-}$ cells, it is possible that Mfn1 and Mfn2 are required for particular but different developmental transitions. As recently shown by Ishihara et al,[48] Drp1$^{-/-}$ mice display severe developmental abnormalities, particularly in the forebrain and die after embryonic day 12.5. In addition, a neural cell-specific Drp1$^{-/-}$ mice die shortly after birth as a result of brain hypoplasia. Thus, it appears that like Opa1 and Mfns, Drp1 and mitochondrial fission are also essential for embryonic development. Supporting this notion, the *heterozygous* dominant-negative mutation of Drp1 in one human subject has been linked to various birth defects, including microcephaly, abnormal brain development, optic atrophy and hypoplasia, lactic acidemia and long-chain fatty acids accumulation, leading to death at 37 days *postpartum*.[49]

Role of Mitochondrial Dynamics in Cell Death

Apoptosis, a genetically driven form of cell death, results in a highly organized dismantling of dying cells. Decomposition of apoptotic cells is mediated by a family of proteases, called caspases that in some modes of apoptosis are regulated by anti- and pro-apoptotic proteins from the BCL-2 family. The main target of the BCL-2 family in mammalian cells is the OMM. BCL-2 proteins can either induce OMM permeabilization and thus apoptosis (e.g., Bax, Bak and Bok) or inhibit it and promote cell survival (e.g., BCL-2, Bcl-xL, Mcl-1). The ratio of pro-apoptotic versus anti-apoptotic BCL-2 family members is a critical factor in regulating susceptibility to cell death and regulation of BCL-2 family proteins, a way to initiate or modulate the apoptotic signal cascade in response to various stimuli, is under stringent control. Various regulatory mechanisms, including regulation of BCL-2 family protein expression, as well as diverse posttranslational modifications regulate the balance of the activities of pro- and anti-apoptotic BCL-2 family proteins in healthy cells and upon induction of apoptosis. Recent reports suggest that in addition to BCL-2 family proteins, factors implicated in the regulation of mitochondrial dynamics are also important for the control of the mitochondrial steps in apoptosis (Table 1).

The discovery that apoptosis might be functionally linked to mitochondrial division, through the activity of Drp1,[50] has stimulated an extensive research effort, capitalizing on the availability of molecular and imaging tools to precisely modulate and monitor mitochondrial network alterations. Although mitochondrial fragmentation induced by certain triggers, including protonophores and elevated Ca^{2+}, does not necessarily lead to activation of the mitochondrial steps in apoptosis, it has been established that mitochondrial fragmentation is a common feature of stress-induced apoptosis in a broad number of cell types and regardless of the apoptotic trigger used. Inhibition of mitochondrial fragmentation in a number of mammalian cell lines by Drp1^{K38A}, a dominant negative mutant of Drp1, resulted in delaying release of cytochrome c from the mitochondria to the cytosol and consequently inhibiting apoptosis.[50] The role for Drp1 in apoptosis execution has also been confirmed by studies using the *C. elegans*[51] and *D. melanogaster*[52] models of developmental cell death, indicating that Drp1 role in cell death is evolutionarily conserved. In *C. elegans* mitochondria fragment in cells that normally undergo programmed cell death during development. This process depends on the BH3-only protein EGL-1 and can be blocked by mutations in the BCL-2-like gene ced-9, indicating that members of the BCL-2 family might function in the regulation of mitochondrial fragmentation

Table 1. *Overview of proteins implicated in regulation of mitochondrial fusion and fission and apoptosis*

Gene Product	Regulated Process	Localization	Loss-of-Function Phenotype	Apoptosis Regulation
Opa1	Fusion	IMM	-Fragmented mitochondria -Decrease in mitochondrial ATP production, $\Delta\Psi_m$ and oxygen consumption	-Downregulation sensitizes cells to apoptosis, -Major role in apoptotic remodeling of mitochondrial cristae
Mitofusins (Mfn1 and Mfn2)	Fusion	OMM	-Fragmented mitochondria - $\Delta\Psi_m$ changes -Decline in mitochondrial function -Developmental defects	-Downregulation sensitizes cells to apoptosis, overexpression delays apoptosis and inhibits Bax oligomerization -Molecular interaction with Bak
Drp1	Fission	OMM, peroxisomes and cytosol	-Elongated mitochondria	-Downregulation, or inhibition delays cytochrome c release and apoptosis, -Direct interaction with Bcl-xL -Stabilized on mitochondria in Bax/Bak dependent manner
Fis1	Fission	OMM and peroxisomes	-Elongated mitochondria	-Downregulation delays apoptosis, overexpression induces cell death, -Molecular interaction with Bcl-xL
Mff	Fission	OMM	-Elongated mitochondria	-Downregulation delays apoptosis
MTP18	Fission	OMM	-Elongated mitochondria	-Downregulation sensitizes cells to apoptosis
Mitofilin	Cristae structure	IMM	-Defective cristae	-Downregulation induces apoptosis
PARL	Cristae structure (processing of Opa1)	IMM	-Defective cristae	-Loss of PARL expression induces apoptosis

in apoptotic cells.[51] However, the study by Breckenridge et al[43] indicated that not apoptosis regulation but elimination of dysfunctional mitochondria might be the main cell death-related function of Drp1 in *C. elegans*. Thus, additional studies are needed to solidify our understanding of molecular mechanism and importance of Drp1 and mitochondrial division for *C. elegans* apoptosis and mitochondrial turnover.

A number of studies have also demonstrated the ability of mitochondrial morphogenesis proteins other than Drp1 to accelerate or delay the apoptotic response of the cell (Table 1).[9,24,33] As in cells with reduced Drp1 activity, inhibition of mitochondrial fission, through silencing of Fis1 and Mff or overexpression of fusion proteins Mfn1 and Mfn2, reduces cell sensitivity to apoptosis and inhibits cytochrome c release. On the other hand, it has been shown that Opa1 downregulation-induced inhibition of mitochondria fusion and subsequent mitochondrial fragmentation leads to unprompted apoptosis and sensitizes cells to various cell death triggers.[9] This is associated with spontaneous release of cytochrome c, $\Delta\Psi_m$ collapse and apoptosis.[9] Based on these studies, one might conclude that mitochondrial fragmentation facilitates cell death and inhibition of this process counteracts apoptosis activation.

Mechanism of the OMM Permeabilization: A Vital Role for Mitochondrial Fusion and Fission?

Despite increasing knowledge about the regulation of mitochondrial fragmentation during cell death, the debate is still open as to how and to what degree alternations in mitochondrial network dynamics participate in the decision phase of mitochondrial apoptosis. For example the mechanism by which stimulated fission of the mitochondria triggers release of cytochrome c is not known. In addition, the effects of mitochondrial fission inhibition on apoptosis progression vary significantly between different published reports. It is also not clear why inhibition of mitochondrial fission has a much stronger effect on cytochrome c release than the release of other apoptotic proteins, e.g., SMAC/Diablo, from the mitochondria to the cytosol. These discrepancies are likely caused by vastly different approaches used to modulate mitochondrial dynamics in these studies, as well as different times of treatments with apoptosis-inducing compounds. Some of the proposed mechanisms of how mitochondrial morphogenesis proteins influence the apoptotic activities of BCL-2 family proteins, including mitochondrial cristae remodeling and direct effect of mitochondrial morphogenesis proteins on the OMM permeabilization, are discussed below.

Opa1 and Cristae Remodeling

It has been proposed that two distinct mechanisms are involved in the efficient release of the apoptotic factors from the mitochondria to the cytosol. One mediates release of cytochrome c, SMAC/Diablo and other proteins across the OMM and another, namely mitochondrial cristae remodeling, is responsible for the redistribution and mobilization of cytochrome c stored in cristae compartments. Cristae remodeling is a process where mitochondrial cristae fuse and the narrow cristae junctions, structures connecting cristae with the IMM, open. This process is thought to enable more efficient release of cytochrome c, likely through the mobilization of this protein in the proximity of the OMM. The "BH3-only" molecule tBid can induce a remodeling of mitochondrial cristae, including the opening of the cristae junctions, leading to mobilization of the cytochrome c stores. Significantly, since the cristae remodeling was detectable in Bax[-/-]/Bak[-/-] mouse embryonic fibroblasts, it has been concluded that the mechanism of cristae remodeling is Bax and Bak and thus the OMM permeabilization, independent. Recent work by Newmeyer and colleagues[53] demonstrated that pharmacological blockade of OMM permeabilization did not prevent cristae remodeling, confirming that mitochondrial cristae remodeling is indeed separable from OMM permeabilization. However, it has been found that activities of Bax and Bak, are required for cristae remodeling and this activity seems to be independent from their role in the OMM permeabilization. Since, Bak can bind Mfn2[54] and in turn Mfn2 directly interacts with Opa1,[11] it is tempting to speculate that Bak (and Bax) are mechanistically linked to cristae remodeling.

As reported by Scorrano and colleagues,[16] proteolysis of Opa1 appears to be a prerequisite for apoptotic cristae changes.[16] Active form of Opa1 prevents cristae opening, perhaps by stabilizing cristae junctions. Opening of cristae junctions is regulated by tBid and correlates with disruption of Opa1 oligomers. Furthermore, during apoptosis "active" forms of Opa1 are diminished in the mitochondria by the concurrent actions of proteolytic enzymes[16] and Opa1 release from the mitochondria to the cytosol.[17,55] Several proteases are involved in the processing of Opa1, including the presinilin-associated rhomboid-like protein (PARL), Yme1, paraplegin and the m-AAA protease complex. The anti-apoptotic form of Opa1 appears to be absent in PARL[-/-] cells, suggesting a critical and specific role for this protease in Opa1-dependent apoptotic cristae remodeling.[16] Furthermore, mammalian cells lacking PARL do not show altered mitochondrial morphology but enhanced ability to release cytochrome c from the IMM to the cytosol.[16] This suggests that PARL might be implicated in the apoptotic Opa1-dependent regulation of cristae morphology, perhaps through processing of a specific, cristae junction-associated subset of Opa1. Consistently, expression of a disassembly-resistant mutant Opa1 (Opa1^{Q297V}) blocked cytochrome c release and apoptosis but not Bax activation.[53] PARL knockout did not affect activation of Bax- and Bak-dependent mitochondrial permeabilization, but increased cell death, further suggesting that alterations in cristae structure, but not OMM permeabilization, are major determinants of abnormal cell death in PARL$^{-/-}$ cells.[16]

In conclusion, the studies discussed above indicate that mitochondrial cristae remodeling contribute to the mitochondrial steps in apoptosis. Yet, Opa1 processing and cristae remodeling in normal cells is already relatively fast. Thus, although Opa1-dependent cristae remodeling appears to modulate cytochrome c release, the significance of this step for the general apoptotic response of the cell needs to be addressed in more detail. It is unclear how the strong pro-apoptotic effect of PARL deficiency could result from the apparently small changes in the cytochrome c release kinetics. Furthermore, the recent work by Sun et al[56] has challenged the importance of cristae remodeling in apoptosis progression. This study revealed that under certain conditions caspase inhibition could block cristae changes without affecting cytochrome c release. Studies of other proteins implicated in the regulation of mitochondrial cristae will undoubtedly help establish more clearly the degree to which cristae remodeling influences apoptosis. For example, Mitofilin, an IMM-associated ~90-kD protein is enriched in the narrow space between the inner boundary and the OMM, where it assembles into large multimeric protein complexes.[57] Down-regulation of Mitofilin in HeLa cells led to decreased cellular proliferation and moderately increased apoptosis. Importantly, ultrastructural EM tomography studies revealed that the IMM failed to form tubular or vesicular cristae and showed up as closely packed stacks of membrane sheets that fused intermittently. It will be interesting to determine whether Mitofilin deficiency-induced cristae alterations contribute to cytochrome c mobilization and apoptosis progression.

Mitochondrial Morphogenesis and the OMM Permeabilization

Despite very extensive effort, the general mechanism of OMM permeabilization and the mechanisms that releases cytochrome c and other factors from the intermembrane space to the cytosol and what kind, if any, of mitochondrial priming that has to occur before cytochrome c is released are not well understood. The published evidence suggests that mitochondrial morphogenesis proteins, perhaps through direct modulation of BCL-2 family proteins, participate in this process. Since mitochondrial morphogenesis proteins, including Mfn2, Fis1 and Drp1, can interact with BCL-2 family proteins,[24,58,59] one could predict that direct molecular interactions between these proteins (e.g., Drp1, Mfn2) and BCL-2 family members take part in the regulation of the mitochondrial steps in apoptosis. Immunoprecipitation studies demonstrated that in healthy cells Mfn1 and Mfn2 interact with Bak,[54] yet upon induction of apoptosis the interaction of Bak with Mfn2 was no longer detectable.[54] On the other hand, molecular interaction of pro-survival Bcl-xL with Drp1 increases in cells committed to death through the treatment with, BH3-peptides mimicking the death domain of BH3-only proteins,[60] suggesting that pro-death signals can stimulate the reorganization of molecular complexes containing both mitochondrial morphogenesis proteins and certain factors from the BCL-2 protein family. This notion is further supported by the data showing that a viral mitochondrial-associated inhibitor of apoptosis (vMIA) can interact with Bax, Bak and Mfn2 and can sequester these proteins in non-active protein complexes.[61,62] Significantly, mitochondrial fusion is also inhibited in vMIA expressing cells. Furthermore, Bax and Bak can influence the assembly of the Mfn2 containing protein complexes and change their submitochondrial distribution and membrane mobility-properties that correlate with different GTP-bound states of Mfn2.[61]

During apoptosis Bax and Bak localize at submitochondrial foci that are also enriched with Drp1,[63,64] suggesting a functional link between these proteins. The expression and likely conformation of Bax and Bak (e.g., pre-apoptotic versus apoptotic) has been also shown to influence trafficking of Drp1 between the cytosol and the OMM.[64] During apoptosis, in synchrony with mitochondrial activation of Bax and before $\Delta\Psi_m$ collapse, Drp1 is stabilized on mitochondrial fission sites. The increased stability of Drp1 at mitochondrial fission sites is promoted by SUMOylation of Drp1 that occurs in a Bax and Bak dependent manner.[64] Since this process does not depend on Fis1, another fission factor, Bax/Bak- dependent regulation of Drp1 in apoptotic cells is likely to be different than the Drp1-mediated fission mechanism in healthy cells. Although it is thought that Fis1 serves as a Drp1 receptor to mediate mitochondrial fission under normal growth condition, Fis1 does not affect stabilization of Drp1 on the mitochondria in apoptotic cells.[64] Mitochondrial fragmentation correlates with apoptotic activation of Bax and Bak and with cytochrome c release induced by these proteins.[7,63] Significantly, the chemical inhibition of Drp1 activity blocked Bid-activated Bax and Bak-dependent cytochrome c release from the mitochondria.[65] Thus, it has been concluded that Drp1 directly regulates mitochondrial permeabilization, independent of Drp1-dependent division of mitochondria.[65] Yet, in contrast to Drp1[-/- 48] or Drp1 RNAi cells,[25,66] chemical inhibition of Drp1 affects release of not only cytochrome c, but also SMAC/Diablo. The reason for this discrepancy is not clear. It needs to be addressed whether Mdivi-1, the inhibitor of Drp1, targets other than Drp1 proteins required for mitochondrial steps in apoptosis, or whether the genetic inhibition of Drp1 results in some compensatory rearrangement within the mitochondria. The anti-apoptotic effect of Mdivi-1 has been recently demonstrated in vivo, in ischemic- and cisplatin-induced mouse models of nephrotoxic renal injury.[67] As determined by histological examination and TUNEL assay, respectively, Mdivi-1 ameliorated tubular damage in renal cortical and outer medulla tissues and inhibited apoptosis. Importantly, mdivi-1 partially suppressed ischemia-induced mitochondrial fragmentation in proximal tubular cells, suggesting that antiapoptotic action of this compound is indeed linked with Drp1-dependent mitochondrial fission.[67]

The significance and mechanism of mitochondrial fission proteins in OMM permeabilization will likely be established in greater detail when more data is available on the role of other fission proteins in this process. For example, it has been shown that Mff, a recently identified fission factor, is also required for cytochrome c release and apoptosis activation.[33] Yet, the mechanism of Mff action has not been reported. In addition, Sugioka et al[68] have demonstrated that Bax and Bak oligomerization, a step in Bax and Bak activation linked to OMM permeabilization, can be inhibited by Mfn1 and Mfn2 overexpression. Thus, considering the direct interaction of Bak and Mfns in healthy cells, the mitochondrial fusion machinery might also be required for the Bax/Bak-dependent steps in apoptosis.

The OMM permeabilization-independent role of Drp1 in the regulation of cell death has recently been reported.[60] It has been shown that BH3 peptides corresponding to the death domain of BH3-only proteins can induce cell death in Bax/Bak DKO cells. Interestingly, this mode of cell death did not include OMM permeabilization and was not inhibited by caspase inhibitors, yet mitochondrial fragmentation and $\Delta\Psi_m$ decreases were detected.[60] Since, BH3 peptides induced an increase in Bcl-xL binding to Drp1, it has been concluded that BH3 peptides bind to pro-survival BCL-2 proteins to engage the Drp1-dependent fission machinery in the absence of Bax and Bak.[60]

Emerging Evidence for the Role of BCL-2 Family Proteins in the Regulation of Mitochondrial Network Dynamics in Healthy Cells

A number of published reports indicate a mechanistic link between mitochondrial fusion and fission regulating proteins (e.g., Drp1, Mfn2 and Opa1) and proteins from the BCL-2 family, not only during apoptosis, but also in healthy cells. The anti-apoptotic BCL-2 family protein Bcl-xL, as well as its *C. elegans* equivalent, localizes to the mitochondria and as was recently shown expression levels of these proteins are important for mitochondrial morphogenesis.[69,70] As shown by Tan et al[70] and confirmed by Breckenrige et al,[71] cells lacking CED-9 have no alteration in mitochondrial size or ultrastructure. Based on this it has been concluded that CED-9 does not regulate mitochondrial dynamics in *C. elegans*.[71] However, CED-9 deficient cells appear more sensitive to mitochondrial fragmentation and increased CED-9 expression in *C. elegans*[70] as well as mammalian cells[72] produces highly interconnected mitochondria, indicating a regulatory role for CED-9 in the regulation of mitochondrial network dynamics. The CED-9 overexpression-induced mitochondrial elongation was partially suppressed by overexpression of DRP-1 and depended on the BH3 binding pocket of CED-9, indicating that in *C. elegans* CED-9 might directly regulate DRP-1.[70] Furthermore, as suggested by immunoprecipitation data showing that CED-9 overexpressed in mammalian cells interacts with Mfn2,[72] CED-9 might regulate both fusion and fission of mitochondria.[72] Consistently, as shown by Berman et al,[69] Bcl-xL a mammalian homologue of CED-9, is implicated in the regulation of both fusion and fission of mitochondrial network in mammalian neurons. It has been demonstrated that in Bcl-xL[−/−] cortical neurons, mitochondrial size was significantly reduced, as compared to wild type cells. Furthermore, as in the case of CED-9 overexpression, overexpression of Bcl-xL induced mitochondrial elongation. The Bcl-xL-induced mitochondrial elongation depended on increased fusion rates of these organelles.[69] These studies also revealed that mitochondrial fission in Bcl-xL overexpressing cells is stimulated, suggesting a role for Bcl-xL in the regulation of both fusion and fission of mitochondria. Furthermore, Bcl-xL controls the mitochondrial biomass and energy-producing capacity of mitochondria under normal growth conditions, before the cell is faced with a life-or-death decision.[69] Thus, it has been concluded that Bcl-xL dependent mitochondrial biogenesis might be critical for balancing mitochondrial network dynamics.[69] BCL-2, another pro-survival BCL-2 family protein, is also required for the regulation of mitochondrial length in Purkinje cell processes.[59] Yet, unlike the case of Bcl-xL, quantitative electron microscopy revealed a significant increase in mitochondrial length in *BCL-2*[−/−] dendrites, as compared to wild type dendrites in the mouse. Mitochondria in *BCL-2*[−/−] mice often contained points where they became constricted, suggests that the elongation of mitochondria might be due to the inhibition or slowdown of the mitochondrial fission process. Based on this it has been proposed that mitochondrial fission occurring during neuronal growth might be critically important for dendrite development and synapse formation and that it is regulated by BCL-2 family protein(s).[59]

In addition to Bcl-xL and BCL-2, Bax and Bak, two pro-apoptotic proteins essential for the execution of an apoptotic signal relayed by other BCL-2 family members, participate in the regulation of mitochondrial network dynamics.[54,61] As in the case of Bcl-xL[−/−] neurons, the reduction of mitochondrial size and lowered fusion rates, have been detected in Bax/Bak double knock out mouse embryonic fibroblasts (Bax/Bak[−/−] MEFs),[61] indicating a role for these proteins in the regulation of mitochondrial fusion. It is not known whether fission of mitochondria in Bax/Bak[−/−] MEFs proceed at the normal rate. However, fragmented mitochondrial in Bax/Bak[−/−] MEFs can be rescued more efficiently by Mfn overexpression-induced stimulation of mitochondrial fusion than by expression of a dominant negative mutant of Drp1 (Drp1[K38A]) and thus by inhibition of mitochondrial division.[61] Therefore, like Bcl-xL and BCL-2, Bax and Bak might also participate in the coordination of both mitochondria fusion and fission.

It needs to be determined how the BCL-2 family proteins intersect mitochondrial fusion and fission. The work by Martin and colleagues[73] indicates that overexpression of Bcl-xL as well as Mcl-1, another anti-apoptotic BCL-2 family protein, can induce mitochondrial fragmentation and that this depends on expression levels of other BCL-2 family proteins. Mcl-1 overexpression-induced mitochondrial fragmentation is dramatically facilitated in Bax/Bak RNAi cells, whereas Bcl-xL overexpression can also fragment mitochondria in cells expressing normal levels of Bax and Bak.[73] These data suggest that the specific activity (e.g., stimulation of fusion or fission) of certain BCL-2 family members in healthy cells is strongly influenced by expression patterns of other BCL-2 family proteins. Therefore, as in case of apoptosis regulation, the overall activity of different members of the BCL-2 family and its integration might be critical for their function in mitochondrial fusion and/or fission. Since pro-apoptotic Bax and Bak and anti-apoptotic Bcl-xL increase mitochondrial length in healthy cells, these proteins might act cooperatively to regulate the steady state of the mitochondrial network prior to apoptosis induction.

Conclusion

Studies indicating that mitochondrial fusion and fission might be vital for apoptosis highlight a critical need for further research of these processes. Although several key fusion and division factors have been identified, the coordination and regulation of these proteins is only preliminarily characterized. The data discussed above link regulation of mitochondrial morphogenesis with mitochondrial steps in apoptosis and vice versa, they also show that BCL-2 family proteins "traditionally" implicated in apoptosis regulation are also important for mitochondrial function in healthy cells. They also reveal how much remains to be determined about the mechanisms by which specific mitochondrial morphogenesis proteins participate in the regulation of apoptosis. The molecular links between BCL-2 family proteins and mitochondrial fusion and fission regulation also needs to be further scrutinized and revealed with more detail.

Addressing these issues could ultimately allow the development of strategies for stabilizing and protecting mitochondria and undoubtedly should result in a number of exciting discoveries bringing us closer to a more complete understanding of mitochondria and their role in cell life and death.

Acknowledgements

I would like to thank P. Wright for comments on the manuscript. I also gratefully acknowledge financial support from National Institute of General Medical Science RO1 GM083131.

References

1. Malka F, Guillery O, Cifuentes-Diaz C et al. Separate fusion of outer and inner mitochondrial membranes. EMBO Rep 2005; 6:853-859.
2. Meeusen S, McCaffery JM, Nunnari J. Mitochondrial fusion intermediates revealed in vitro. Science 2004; 305:1747-1752.
3. Chen H, Detmer SA, Ewald AJ et al. Mitofusins Mfn1 and Mfn2 coordinately regulate mitochondrial fusion and are essential for embryonic development. J Cell Biol 2003; 160:189-200.
4. Ishihara N, Eura Y, Mihara K. Mitofusin 1 and 2 play distinct roles in mitochondrial fusion reactions via GTPase activity. J Cell Sci 2004; 117:6535-6546.
5. Koshiba T, Detmer SA, Kaiser JT et al. Structural basis of mitochondrial tethering by mitofusin complexes. Science 2004; 305:858-862.
6. Neuspiel M, Zunino R, Gangaraju S et al. Activated mitofusin 2 signals mitochondrial fusion, interferes with Bax activation and reduces susceptibility to radical induced depolarization. J Biol Chem 2005; 280:25060-25070.
7. Karbowski M, Arnoult D, Chen H et al. Quantitation of mitochondrial dynamics by photolabeling of individual organelles shows that mitochondrial fusion is blocked during the Bax activation phase of apoptosis. J Cell Biol 2004; 164:493-499.
8. Delettre C, Lenaers G, Griffoin JM et al. Nuclear gene OPA1, encoding a mitochondrial dynamin-related protein, is mutated in dominant optic atrophy. Nat Genet 2000; 26:207-210.
9. Olichon A, Baricault L, Gas N et al. Loss of OPA1 perturbates the mitochondrial inner membrane structure and integrity, leading to cytochrome c release and apoptosis. J Biol Chem 2003; 278:7743-7746.
10. Detmer SA, Chan DC. Functions and dysfunctions of mitochondrial dynamics. Nat Rev Mol Cell Biol 2007; 8:870-879.
11. Guillery O, Malka F, Landes T et al. Metalloprotease-mediated OPA1 processing is modulated by the mitochondrial membrane potential. Biol Cell 2008; 100:315-325.
12. Eura Y, Ishihara N, Oka T et al. Identification of a novel protein that regulates mitochondrial fusion by modulating mitofusin (Mfn) protein function. J Cell Sci 2006; 119:4913-4925.
13. Choi SY, Huang P, Jenkins GM et al. A common lipid links Mfn-mediated mitochondrial fusion and SNARE-regulated exocytosis. Nat Cell Biol 2006; 8:1255-1262.
14. Hajek P, Chomyn A, Attardi G. Identification of a novel mitochondrial complex containing mitofusin 2 and stomatin-like protein 2. J Biol Chem 2007; 282:5670-5681.
15. Da Cruz S, Parone PA, Gonzalo P et al. SLP-2 interacts with prohibitins in the mitochondrial inner membrane and contributes to their stability. Biochim Biophys Acta 2008; 1783:904-911.
16. Cipolat S, Rudka T, Hartmann D et al. Mitochondrial rhomboid PARL regulates cytochrome c release during apoptosis via OPA1-dependent cristae remodeling. Cell 2006; 126:163-175.

17. Griparic L, Kanazawa T, van der Bliek AM. Regulation of the mitochondrial dynamin-like protein Opa1 by proteolytic cleavage. J Cell Biol 2007; 178:757-764.
18. Song Z, Chen H, Fiket M et al. OPA1 processing controls mitochondrial fusion and is regulated by mRNA splicing, membrane potential and Yme1L. J Cell Biol 2007; 178:749-755.
19. Ishihara N, Fujita Y, Oka T et al. Regulation of mitochondrial morphology through proteolytic cleavage of OPA1. EMBO J 2006; 25:2966-2977.
20. Duvezin-Caubet S, Koppen M, Wagener J et al. OPA1 processing reconstituted in yeast depends on the subunit composition of the m-AAA protease in mitochondria. Mol Biol Cell 2007; 18:3582-3590.
21. Smirnova E, Griparic L, Shurland DL et al. Dynamin-related protein Drp1 is required for mitochondrial division in mammalian cells. Mol Biol Cell 2001; 12:2245-2256.
22. Stojanovski D, Koutsopoulos OS, Okamoto K et al. Levels of human Fis1 at the mitochondrial outer membrane regulate mitochondrial morphology. J Cell Sci 2004; 117:1201-1210.
23. Yoon Y, Krueger EW, Oswald BJ et al. The mitochondrial protein hFis1 regulates mitochondrial fission in mammalian cells through an interaction with the dynamin-like protein DLP1. Mol Cell Biol 2003; 23:5409-5420.
24. James DI, Parone PA, Mattenberger Y et al. hFis1, a novel component of the mammalian mitochondrial fission machinery. J Biol Chem 2003; 278:36373-36379.
25. Lee YJ, Jeong SY, Karbowski M et al. Roles of the mammalian mitochondrial fission and fusion mediators Fis1, Drp1 and Opa1 in apoptosis. Mol Biol Cell 2004; 15:5001-5011.
26. Chang CR, Blackstone C. Drp1 phosphorylation and mitochondrial regulation. EMBO Rep 2007; 8:1088-1089; author reply 1089-1090.
27. Harder Z, Zunino R, McBride H. Sumo1 conjugates mitochondrial substrates and participates in mitochondrial fission. Curr Biol 2004; 14:340-345.
28. Karbowski M, Neutzner A, Youle RJ. The mitochondrial E3 ubiquitin ligase MARCH5 is required for Drp1 dependent mitochondrial division. J Cell Biol 2007; 178:71-84.
29. Nakamura N, Hirose S. Regulation of mitochondrial morphology by USP30, a deubiquitinating enzyme present in the mitochondrial outer membrane. Mol Biol Cell 2008; 19:1903-1911.
30. Cribbs JT, Strack S. Reversible phosphorylation of Drp1 by cyclic AMP-dependent protein kinase and calcineurin regulates mitochondrial fission and cell death. EMBO Rep 2007; 8:939-944.
31. Han XJ, Lu YF, Li SA et al. CaM kinase I alpha-induced phosphorylation of Drp1 regulates mitochondrial morphology. J Cell Biol 2008; 182:573-585.
32. Taguchi N, Ishihara N, Jofuku A et al. Mitotic phosphorylation of dynamin-related GTPase Drp1 participates in mitochondrial fission. J Biol Chem 2007; 282:11521-11529.
33. Gandre-Babbe S, van der Bliek AM. The novel tail-anchored membrane protein Mff controls mitochondrial and peroxisomal fission in mammalian cells. Mol Biol Cell 2008; 19:2402-2412.
34. Yonashiro R, Ishido S, Kyo S et al. A novel mitochondrial ubiquitin ligase plays a critical role in mitochondrial dynamics. EMBO J 2006; 25:3618-3626.
35. Nakamura N, Kimura Y, Tokuda M et al. MARCH-V is a novel mitofusin 2- and Drp1-binding protein able to change mitochondrial morphology. EMBO Rep 2006; 7:1019-1022.
36. Braschi E, Zunino R, McBride HM. MAPL is a new mitochondrial SUMO E3 ligase that regulates mitochondrial fission. EMBO Rep 2009; 10:748-754.
37. Zunino R, Schauss A, Rippstein P et al. The SUMO protease SENP5 is required to maintain mitochondrial morphology and function. J Cell Sci 2007; 120:1178-1188.
38. Poole AC, Thomas RE, Andrews LA et al. The PINK1/Parkin pathway regulates mitochondrial morphology. Proc Natl Acad Sci USA 2008; 105:1638-1643.
39. Yang Y, Ouyang Y, Yang L et al. Pink1 regulates mitochondrial dynamics through interaction with the fission/fusion machinery. Proc Natl Acad Sci USA 2008; 105:7070-7075.
40. Nakada K, Inoue K, Ono T et al. Inter-mitochondrial complementation: Mitochondria-specific system preventing mice from expression of disease phenotypes by mutant mtDNA. Nat Med 2001; 7:934-940.
41. Varadi A, Johnson-Cadwell LI, Cirulli V et al. Cytoplasmic dynein regulates the subcellular distribution of mitochondria by controlling the recruitment of the fission factor dynamin-related protein-1. J Cell Sci 2004; 117:4389-4400.
42. Twig G, Elorza A, Molina AJ et al. Fission and selective fusion govern mitochondrial segregation and elimination by autophagy. EMBO J 2008; 27:433-446.
43. Breckenridge DG, Kang BH, Kokel D et al. Caenorhabditis elegans drp-1 and fis-2 regulate distinct cell-death execution pathways downstream of ced-3 and independent of ced-9. Mol Cell 2008; 31:586-597.
44. Benard G, Bellance N, James D et al. Mitochondrial bioenergetics and structural network organization. J Cell Sci 2007; 120:838-848.

45. Chen H, Chomyn A, Chan DC. Disruption of fusion results in mitochondrial heterogeneity and dysfunction. J Biol Chem 2005; 280:26185-26192.
46. Tondera D, Grandemange S, Jourdain A et al. SLP-2 is required for stress-induced mitochondrial hyperfusion. EMBO J 2009; 28:1589-1600.
47. Davies VJ, Hollins AJ, Piechota MJ et al. Opa1 deficiency in a mouse model of autosomal dominant optic atrophy impairs mitochondrial morphology, optic nerve structure and visual function. Hum Mol Genet 2007; 16:1307-1318.
48. Ishihara N, Nomura M, Jofuku A et al. Mitochondrial fission factor Drp1 is essential for embryonic development and synapse formation in mice. Nat Cell Biol 2009, doi:10.1038/ncb1907.
49. Waterham HR, Koster J, van Roermund CW et al. A lethal defect of mitochondrial and peroxisomal fission. N Engl J Med 2007; 356:1736-1741.
50. Frank S, Gaume B, Bergmann-Leitner ES et al. The role of dynamin-related protein 1, a mediator of mitochondrial fission, in apoptosis. Dev Cell 2001; 1:515-525.
51. Jagasia R, Grote P, Westermann B et al. DRP-1-mediated mitochondrial fragmentation during EGL-1-induced cell death in C. elegans. Nature 2005; 433:754-760.
52. Goyal G, Fell B, Sarin A et al. Role of mitochondrial remodeling in programmed cell death in Drosophila melanogaster. Dev Cell 2007; 12:807-816.
53. Yamaguchi R, Lartigue L, Perkins G et al. Opa1-mediated cristae opening is Bax/Bak and BH3 dependent, required for apoptosis and independent of Bak oligomerization. Mol Cell 2008; 31:557-569.
54. Brooks C, Wei Q, Feng L et al. Bak regulates mitochondrial morphology and pathology during apoptosis by interacting with mitofusins. Proc Natl Acad Sci USA 2007; 104:11649-11654.
55. Arnoult D, Gaume B, Karbowski M et al. Mitochondrial release of AIF and EndoG requires caspase activation downstream of Bax/Bak-mediated permeabilization. EMBO J 2003; 22:4385-4399.
56. Sun MG, Williams J, Munoz-Pinedo C et al. Correlated three-dimensional light and electron microscopy reveals transformation of mitochondria during apoptosis. Nat Cell Biol 2007; 9:1057-1065.
57. John GB, Shang Y, Li L et al. The mitochondrial inner membrane protein mitofilin controls cristae morphology. Mol Biol Cell 2005; 16:1543-1554.
58. Li H, Chen Y, Jones AF et al. Bcl-xL induces Drp1-dependent synapse formation in cultured hippocampal neurons. Proc Natl Acad Sci USA 2008; 105:2169-2174.
59. Liu QA, Shio H. Mitochondrial morphogenesis, dendrite development and synapse formation in cerebellum require both BCL-2 and the glutamate receptor delta2. PLoS Genet 2008; 4:e1000097.
60. Shroff EH, Snyder CM, Budinger GR et al. BH3 peptides induce mitochondrial fission and cell death independent of BAX/BAK. PLoS One 2009; 4:e5646.
61. Karbowski M, Norris KL, Cleland MM et al. Role of Bax and Bak in mitochondrial morphogenesis. Nature 2006; 443:658-662.
62. Norris KL, Youle RJ. Cytomegalovirus proteins vMIA and m38.5 link mitochondrial morphogenesis to BCL-2 family proteins. J Virol 2008; 82:6232-6243.
63. Karbowski M, Lee YJ, Gaume B et al. Spatial and temporal association of Bax with mitochondrial fission sites, Drp1 and Mfn2 during apoptosis. J Cell Biol 2002; 159:931-938.
64. Wasiak S, Zunino R, McBride HM: Bax/Bak promote sumoylation of DRP1 and its stable association with mitochondria during apoptotic cell death. J Cell Biol 2007; 177:439-450.
65. Cassidy-Stone A, Chipuk JE, Ingerman E et al. Chemical inhibition of the mitochondrial division dynamin reveals its role in Bax/Bak-dependent mitochondrial outer membrane permeabilization. Dev Cell 2008; 14:193-204.
66. Parone PA, James DI, Da Cruz S et al. Inhibiting the mitochondrial fission machinery does not prevent Bax/Bak-dependent apoptosis. Mol Cell Biol 2006; 26:7397-7408.
67. Brooks C, Wei Q, Cho SG et al. Regulation of mitochondrial dynamics in acute kidney injury in cell culture and rodent models. J Clin Invest 2009; 119:1275-1285.
68. Sugioka R, Shimizu S, Tsujimoto Y. Fzo1, a protein involved in mitochondrial fusion, inhibits apoptosis. J Biol Chem 2004; 279:52726-52734.
69. Berman SB, Chen Y-b, Qi B et al. Bcl-xL increases mitochondrial fission, fusion and biomass in neurons. J Cell Biol 2009; 184:10.1083/jcb.200809060.
70. Tan FJ, Husain M, Manlandro CM et al. CED-9 and mitochondrial homeostasis in C. elegans muscle. J Cell Sci 2008; 121:3373-3382.
71. Breckenridge DG, Kang BH, Xue D. BCL-2 proteins EGL-1 and CED-9 do not regulate mitochondrial fission or fusion in Caenorhabditis elegans. Curr Biol 2009; 19:768-773.
72. Delivani P, Adrain C, Taylor RC et al. Role for CED-9 and Egl-1 as regulators of mitochondrial fission and fusion. Mol Cell 2006; 21:761-773.
73. Sheridan C, Delivani P, Cullen SP et al. Bax- or Bak-induced mitochondrial fission can be uncoupled from cytochrome C release. Mol Cell 2008; 31:570-585.

INDEX